AF576653

Vermeer's Camera

Philip Steadman is Professor of Urban and Built Form Studies at University College London. He trained as an architect, and has taught at Cambridge University and the Open University. He has published several books on geometry in architecture, and on computer-aided design. In the 1960s he edited and published *Form*, a quarterly magazine of the arts, and co-authored a book on kinetic art. He helped to produce four computer-animated films on the work of Leonardo da Vinci for an exhibition in London in 1989. He has also contributed to other exhibitions, films, and books on perspective geometry and the history of art. *Vermeer's Camera* is the product of twenty years' fascination with the Dutch painter.

Vermeer's Camera

Uncovering the truth behind the masterpieces

Philip Steadman

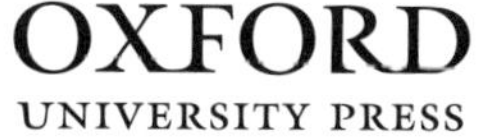

UNIVERSITY PRESS

OXFORD
UNIVERSITY PRESS

Great Clarendon Street, Oxford OX2 6DP

Oxford University Press is a department of the University of Oxford.
It furthers the University's objective of excellence in research, scholarship,
and education by publishing worldwide in

Oxford New York

Auckland Bangkok Buenos Aires Cape Town Chennai
Dar es Salaam Delhi Hong Kong Istanbul Karachi Kolkata
Kuala Lumpur Madrid Melbourne Mexico City Mumbai Nairobi
São Paulo Shanghai Singapore Taipei Tokyo Toronto

and an associated company in Berlin

Oxford is a registered trade mark of Oxford University Press
in the UK and in certain other countries

Published in the United States
by Oxford University Press Inc., New York

© Philip Steadman 2001

The moral rights of the author have been asserted
Database right Oxford University Press (maker)

First published 2001
First published as an Oxford University Press paperback 2002

All rights reserved. No part of this publication may be reproduced,
stored in a retrieval system, or transmitted, in any form or by any means,
without the prior permission in writing of Oxford University Press,
or as expressly permitted by law, or under terms agreed with the appropriate
reprographics rights organization. Enquiries concerning reproduction
outside the scope of the above should be sent to the Rights Department,
Oxford University Press, at the address above

You must not circulate this book in any other binding or cover
and you must impose this same condition on any acquirer

British Library Cataloguing in Publication Data
Data available

Library of Congress Cataloging in Publication Data
Data available

ISBN 0-19-280302-6

1 3 5 7 9 10 8 6 4 2

Typeset in Palatino and Stone Sans
by RefineCatch Limited, Bungay, Suffolk
Printed in Spain by Book Print S. L.

CONTENTS

ACKNOWLEDGEMENTS

This book has been a long time in the making. It has its origins in an exercise in perspective analysis set to summer school students in the 1970s, in which we reconstructed the three-dimensional spaces shown in Dutch 17th-century genre paintings. Many of these pictures are specially suited for such an exercise, because Dutch houses of the period have tiled floors. This started me wondering about whether Vermeer's interiors showed the same or different rooms, so I set about making reconstructions for myself (Chapter 5). At different times my friend and colleague Frank Brown helped with the perspective analyses, and Rachel Hewitt made bird's-eye views. The Northampton firm of Bassett Lowke built a beautiful model of the room in hardwood and brass (Chapter 7), for which the Open University generously met the costs. George Beale, who previously worked as a cabinet-maker in High Wycombe, built some of the miniature furniture. Cyril Hughes used all his ingenuity and skill in manufacturing other furnishings and fittings. Sue Read sewed costumes for the figures in *The Music Lesson*. Deborah Hopson made tiny Delftware jugs.

The first photographs of the model, which appear here as Plates 5, 7, and 9 and Figures 60 and 61, were taken in 1982/3 by my good friend and neighbour Trevor Yorke. He and I still remember the thrill of seeing our first simulated Vermeers looming out of the whiteness of the photographic paper in the developing tray. Trevor's professionalism and enthusiasm were equalled by Richard Hearne at the Open University, who carried on the photographic work at a later stage, and took the pictures which appear here as Figures 62 to 64.

In 1989 Hendrik Ball and George Auckland devoted an edition of their popular science television programme *Take Nobody's Word for It* to Vermeer and the camera obscura. John Bone built a full-size replica of the room in the BBC's Bristol studios, dressed as for *The Music Lesson*. The parts of the woman at the virginals and her admiring teacher were played by Carol Vorderman and Ian Fells. (I played the artist.) The set and camera obscura constructed for this film proved the feasibility of projecting optical images at the size of Vermeer's originals, indoors, with a single simple lens (Chapter 7). At another time Karl Sabbagh also planned a film on Vermeer and the camera obscura, which was never made; however, his research has been useful for this book, especially his material on Torrentius.

I have been fortunate to receive generous support and advice from two distinguished academic colleagues. The art historian Martin Kemp saw the scale model of Vermeer's room and was kind enough to devote a page of his encyclopaedic *The Science of Art* (1990) to photographs and a description—the work's first appearance in print.[1] The psychologist Richard Gregory visited the set of the 1989 film, and invited me to contribute an essay to *The Artful Eye* (1995), which he edited with John Harris, Priscilla Heard, and David Rose.[2]

I have also been helped by several Vermeer scholars, Vermeer enthusiasts, and specialists in optics, in correspondence over the years. I am grateful to John Michael Montias for information about Maria Thins's house on the Molenpoort and other topographical details. My friend Marc van Leusen kindly made measurements of furniture in the Prinsenhof Museum in Delft. Catherine Cooke and Cyril Hughes measured buildings and took photographs in Delft. The staff of the Rijksmuseum in Amsterdam and the Municipal Archives in Delft have been both friendly and efficient. The cartographic department of the Technological University located an 1830 map of Delft that has proved useful in several ways.

Gillis van Oosten generously sent me local information about his home-town including a copy of Michel van Maarseveen's book *Vermeer of Delft*.[3] Tracy Chevalier presented me with her facsimile copy of the 1675–8 *Kaart Figuratief van Delft*.[4] Allan Mills of the University of Leicester Astronomy Group sent me papers about 17th-century lenses—although he might not perhaps agree with all my conclusions here. John Sharp and Fred Dubery supplied articles about Vermeer. Deirdre Bethune made translations from the Dutch. Ruth Brandon read the manuscript and made several suggestions. Judith Field did the same and saved me from many solecisms in the history of optics. Finally I want to express special thanks for the long correspondence, and the most stimulating and suggestive discussions, which I have had over several years with Quentin Williams. I am especially grateful to Quentin for reading an earlier draft of this book and making many useful and provocative comments.

LIST OF PLATES

1 Vermeer, *Officer and Laughing Girl*, *c*.1658. Oil on canvas, 50.5 × 46 cm.

2 Vermeer, *Girl with a Red Hat*, *c*.1660–1. Oil on wood panel, 23 × 18 cm.

3 Vermeer, *View of Delft*, *c*.1660–1. Oil on canvas, 98.5 × 117.5 cm.

4 Vermeer, *The Geographer*, *c*.1668–9. Oil on canvas, 53 × 46.6 cm.

5 and 6 Photographic reconstruction of *The Music Lesson* (top), compared with the real painting (bottom). Oil on canvas, 73.3 × 64.5 cm.

7 and 8 Photographic reconstruction of *The Concert* (left), compared with the real painting (right). Oil on canvas, 72.5 × 64.7 cm.

9 and 10 Photographic reconstruction of *Lady Standing at the Virginals* (top), compared with the real painting (bottom). Oil on canvas, 51.7 × 45.2 cm.

LIST OF ILLUSTRATIONS

INTRODUCTION

In this book I seek to show precisely how the Dutch artist Johannes Vermeer (1632–75) used the camera obscura as an aid to painting. The camera obscura was the predecessor of the photographic camera. It is a simple device incorporating a pinhole or lens, with which an image of a scene can be projected onto a screen. The image can then be traced.

The idea that Vermeer might have made occasional use of some kind of optical instrument to support or inspire his compositional and painting technique is one that has become widely held among art historians, as we shall see. Where scholars have had more difficulty is in accepting the proposition that an artist with such outstanding technical skills as Vermeer might have used the camera wholesale, to trace large parts of the outlines of many paintings. This reluctance is doubtless related to old and—one might have thought—long-since settled controversies about whether photography can be an art. Copying mechanically produced images seems, by pre-modernist standards of artistic craftmanship, a dubious and underhand practice, to which only the incompetent would resort. As Martin Kemp puts it, in a comprehensive account of artificial aids to painting in 17th-century Dutch art:

> Art historians have generally been reluctant to study the implications of this evidence, feeling, no doubt, that it is not quite proper for their favoured artists to resort to what has become regarded as a form of cheating.[1]

I believe that reservations of this kind are misplaced and historically irrelevant in the case of Vermeer. The camera allowed the artist to enter a newly revealed world of optical phenomena and to explore how these might be recorded in paint. It is important to recognize that in the mid-17th century the very idea of producing pictorial images with lenses was both novel and prestigious. Comparisons with the instantaneity of modern photography are in any case misleading. The use of the camera obscura for painting is hardly a matter of short cuts or technical ease. On the contrary, it forces protracted and attentive looking and analysis. What is more, the camera obscura, when used to capture subjects like domestic interiors, does not impose a given compositional structure on the artist. On the contrary, it can be used as a device to aid the very process of composition. Sitters and furniture can be positioned and adjusted and the effects on the resulting two-dimensional image judged. As Kemp puts it, 'the use of a camera in no way prescribes the artistic choices to be made at each stage of the conception and making of a painting.'[2]

Indeed I suggest that Vermeer's obsessions with light, tonal values, shadow, and colour, for the treatment of which his work is so much admired, are very closely bound up with his study of the special qualities of optical images. In arguing this I will lean heavily on what is probably the subtlest and most sensitive of all analyses of Vermeer's technique, that of Lawrence Gowing—himself a painter—in his beautiful monograph of 1952.[3] Gowing shares the general view 'that Vermeer made use of the camera obscura.'[4] Unlike other critics, however, Gowing goes on to claim that 'Vermeer is alone in putting it to the service of style rather than the accumulation of facts.' What Gowing calls the painter's 'explanatory vocabulary', his 'interruption and denial of line', his 'optical impartiality', and above all the 'unvarying adequacy, the uniform success of [his] method'—all these can be attributed, as Gowing argues, to a technique which depended on careful, prolonged observation of patterns of light falling on the camera screen.

Previous theorists have drawn their evidence for Vermeer's use of the camera from certain idiosyncrasies of his painted treatment which seem to mimic the distortions produced by lenses. Thus Vermeer seems to render some passages 'out of focus', and elsewhere seems to reproduce other effects which would be seen through the camera but not with the naked eye. The evidence presented here is different. It relies on an analysis of the perspective geometry of the paintings.

The supporting detail is complex, but the central core of the argument is simple enough: Vermeer seems to have painted as many as a dozen pictures in the same room. Because he constructs his perspective views—by whatever method—with such precision, it is possible to measure the shape and dimensions of this room to a high degree of accuracy. The precise position of the theoretical viewpoint—the notional point in space at which Vermeer placed his eye—can be determined in each case. Everything which is visible in the painting itself must be contained within a 'visual pyramid', whose apex is at this viewpoint. It is possible to determine the positions of the sloping lines that form the edges of this pyramid. Suppose that these lines are continued back, through the viewpoint, to meet the back wall of the room, behind the painter. They then define a rectangular area on that wall. In at least half a dozen cases this rectangle is the precise size of the painting in question.

My explanation for this very curious result is that Vermeer had a camera obscura with a lens at the painting's viewpoint. He used this arrangement to project the scene onto the back wall of the room, which thus served as the camera's screen. He put paper on the wall and traced, perhaps even painted from the projected image. It is because Vermeer traced the images that they are the same size as the paintings themselves.

I argue that it is difficult to account for this geometrical phenomenon by reference to other procedures which Vermeer might plausibly or feasibly have adopted: using conventional mathematical methods to set up the perspective views, for example, or tracing the images reflected in mirrors, as more than one critic has suggested. By contrast the result is very simply and naturally explicable, once it is supposed that Vermeer was using a camera. The geometrical nature of this argument gives it a special force; if the camera theory is to be rejected, a viable alternative explanation still needs to be developed for a most unusual property of the perspective construction of Vermeer's interiors.

Lawrence Gowing wrote of the problem of the optical basis of Vermeer's technique, 'the precise technical solution remains a matter of conjecture. The truth is buried.'[5] I hope to have disinterred at least part of the truth and in so doing to have avoided something that Gowing feared: using the pictures 'no better than as fodder for another of the hobby-horses to which more than one study of the painter has been bound.' My purpose has definitely not been to make 'a study of the painter'. In no way is the book intended as a comprehensive technical or critical treatment of Vermeer's *oeuvre*. I have concentrated on just one aspect of Vermeer's working methods, but one which has an intrinsic fascination in itself and which furthermore concerns the painter's special genius for capturing the qualities of light.

1 The camera obscura

If a small hole is made in the wall of a darkened room, an image of the scene outside can be formed by light rays passing through the hole. The image may appear on a wall opposite the hole, or can be observed on a sheet of paper or other screen placed in front of the hole. The hole can be in a door, say, or in a solid wooden window shutter. This is the 'camera obscura' in its original meaning,[1] the term coming from the astronomer Kepler in the early 17th century.[2]

Typically the image in this kind of 'pinhole' camera obscura is very dim. ('Pinhole' is perhaps misleading. In a room-sized camera the aperture might be the size of a keyhole.) For the image to be sharp, the hole must be small, and this of course limits the quantity of light admitted. The outside scene must therefore be very brightly lit—but even the image of a landscape under the Mediterranean summer sun is still faint. A person entering the darkened room must wait a few minutes for his eyes to adjust, before he can make out the picture.

One exception is an image of the sun itself, which is immediately and clearly visible. The effect can be seen in attics under ill-fitting tiled roofs, where the sun's rays are focused through chinks between the tiles. The same can happen when sunlight penetrates thin canopies of leaves. Aristotle watched an image of the sun formed beneath a tree becoming crescent-shaped during an eclipse.[3] In this he anticipated the main practical application of the camera obscura up until the 17th century, which was for watching solar eclipses without damaging one's eyes.

The question of how images are formed by light as it passes through small apertures was studied by Chinese philosophers in the 8th century (and possibly even earlier), and by the great scholar Alhazen writing in Arabic in the 10th century.[4] Alhazen's work on optics exerted a powerful influence on some of the most important 13th-century writers on the subject, the English Franciscans Roger Bacon and John Pecham and the Polish philosopher Witelo.[5] These three were in turn to dominate the study of optics by their writings for another two hundred years, until the time of Kepler. For example, in his *Perspectiva Communis* (*Natural Optics*) of 1279, Pecham describes how 'When at the time of an eclipse of the sun, its rays are received in a dark place through a hole of any shape, it is possible to see the crescent-shape getting smaller as the moon covers the sun.'[6] By the mid-15th century the use of the darkened-room type of camera obscura for making solar observations had become standard practice among astronomers. The first published illustration of a camera (of this or any kind) appears in a book by the Dutch mathematician and astronomer Reinerus Gemma Frisius (Figure 1).[7] It shows an image of the solar eclipse which Gemma Frisius observed at Louvain in 1544, projected onto the wall of a classical pavilion.

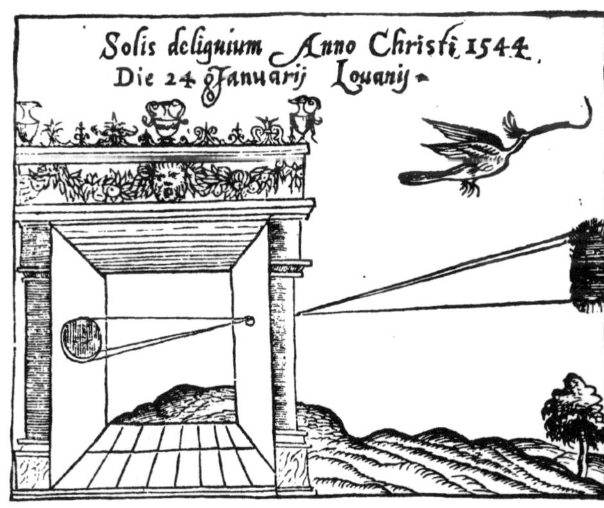

Figure 1
Eclipse of the sun observed in a camera obscura by Gemma Frisius at Louvain in 1544; from *De Radio Astronomico et Geometrico* (2nd edn 1558). See how the image of the sun is inverted.

For other 16th- and 17th-century authors the camera was merely a curious device for spying on people in the street outside—a trick from the repertoire of 'natural magic'. The first hints of possible uses of the camera obscura in art come in Leonardo da Vinci's notebooks, around the year 1490:

> An experiment, showing how objects transmit their images or pictures intersecting within the eye in the crystalline humour.
>
> This is shown when the images of illuminated objects penetrate into a very dark chamber by some small round hole. Then you will receive these images on a white paper placed within this dark room rather near to the hole; and you will see all the objects on the paper in their proper forms and colours, but much smaller; and they will be upside down by reason of that very intersection. These images, being transmitted from a place illuminated by the sun, will seem as if actually painted on this paper, which must be extremely thin and looked at from behind. And let the little perforation be made in a very thin plate of iron.[8]

Leonardo remarks on how the projected image is inverted in relation to the original scene, since the light rays cross as they pass through the pinhole. If a camera image is studied from the direction that the light comes—as in the illustration of Gemma Frisius—then it appears both upside down and reversed left-to-right (Figure 2a). One further problem here is that the viewer's head can get in the way of the light. To overcome this second difficulty Leonardo proposes a translucent paper screen, viewed from the *back*. The image as the viewer sees it is still upside down, but it is *not* now reversed left-to-right (Figure 2b). There is, however, some loss of brightness as the light passes through the paper.

An important technical development in the camera obscura which took place in the middle of the 16th century was the introduction of glass lenses in the place of simple pinholes. This made it possible to have larger apertures and hence much brighter images, without sacrificing sharpness. As a result it became a real practical proposition to use the apparatus for making drawings from life. Convex lenses suitable for the purpose had been widely available for several centuries. The first use of spectacles for correcting long-sightedness has been traced to Italy, towards the end of the 13th century. By the 16th century spectacles were being produced in quantity and the manufacture of lenses had become an industry.[9]

There is some doubt about which writer deserves credit for the earliest description of a camera with a lens. The honour should possibly go to the Milanese physician and natural philosopher Girolamo Cardano (now best remembered as a mathematician). The relevant passage from his *De Subtilitate* (*On Subtlety*, 1550) runs as follows:

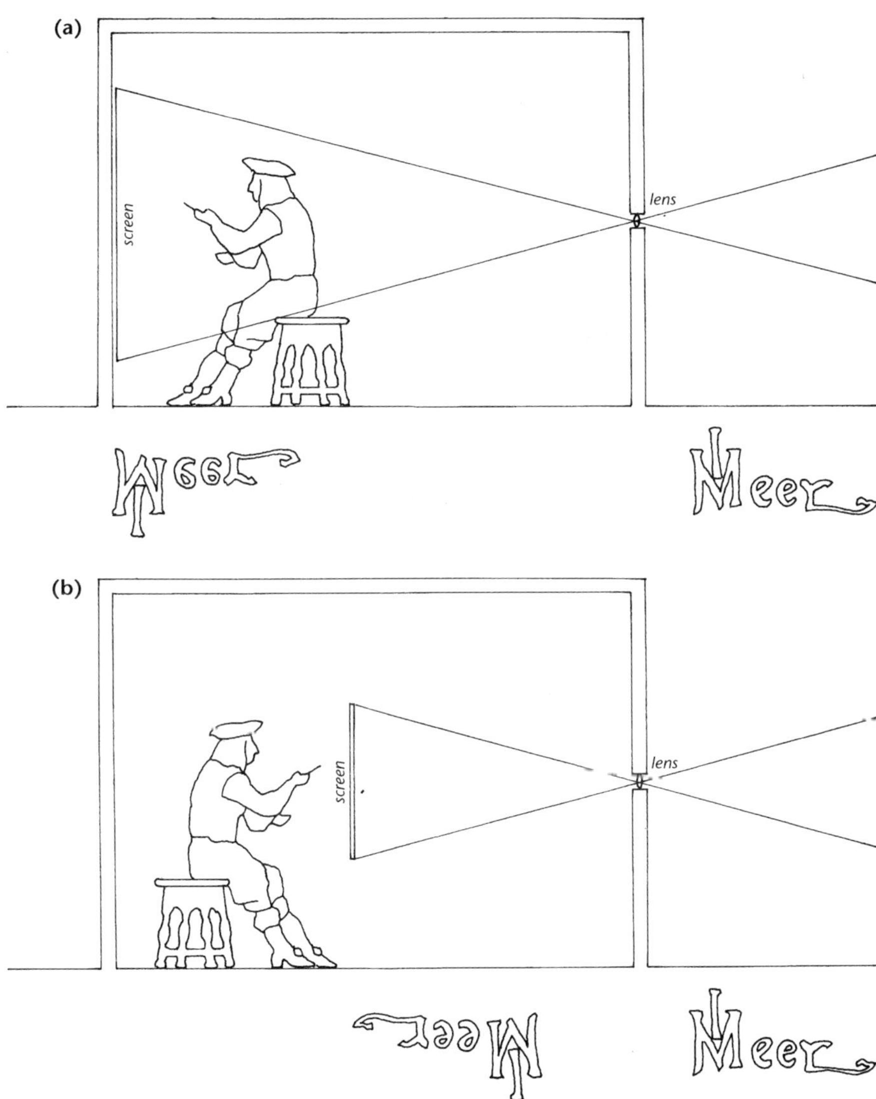

Figure 2 Diagrams of room-type camera obscuras (a) with the image projected onto the far wall and (b) with the image projected onto a translucent screen, viewed from the back. The versions of Vermeer's monogram show the effect of each type of camera on the orientation of the image, as the viewer sees it. In (a) the image is upside-down and reversed left-to-right. In (b) the image is upside-down but not laterally reversed.

> If you care to see what goes on in the street when the sun is bright, place in your windows a glass disc and the window having been closed [shuttered] you will see images projected through the aperture onto the wall; but the colours are dull.[10]

The problem is what does Cardano mean here by 'a glass disc' [in the original Latin: *orbis e vitra*]? Some historians of photography have concluded that it must indeed be a lens.[11] But Major-General Waterhouse who compiled a very learned set of 'Notes on the early history of the camera obscura' suggested that Cardano might have been referring to a concave *mirror*, used to reflect and concentrate light from the outdoor scene onto the pinhole—something described previously by a number of other writers.[12] A lens should certainly not produce colours that are 'dull': 'intense' and 'jewel-like' are adjectives more often used.

Two unequivocal descriptions of cameras with convex lenses appear very soon afterwards, the first in a perspective manual published in 1568 by Daniele Barbaro, a Venetian patrician famous for his editions of the Roman architectural writer Vitruvius.[13] Barbaro proposes explicitly that the camera be used for producing drawings in correct perspective. He says that an old man's spectacle glass (i.e. a convex lens) is suitable. He suggests that a diaphragm may be placed over the lens to restrict the aperture, and so increase the depth of field—although this reduces brightness. He also describes how 'Seeing, therefore, on the paper the outline of things, you can draw with a pencil all the perspective and the shading and colouring, according to nature, holding the paper tightly till you have finished the drawing.'[14]

A second Venetian author Giovanni Battista Benedetti, writing in 1585, suggested a method for correcting the inversion of the image, by setting a plane mirror at 45 degrees to the direction of the light coming from the lens.[15] This is essentially the arrangement used in most of the portable box-type cameras manufactured from the late 17th century onwards.[16] These small instruments were the forerunners of the photographic camera; indeed they have the same arrangement as the modern reflex camera. Figure 3 shows the principle diagrammatically and Figure 4 shows a 19th-century instrument of this type. The image is projected upwards onto a translucent screen, typically of ground glass—although other translucent materials can also serve. When viewed from above, the image appears the correct way up (but still reversed left-to-right).

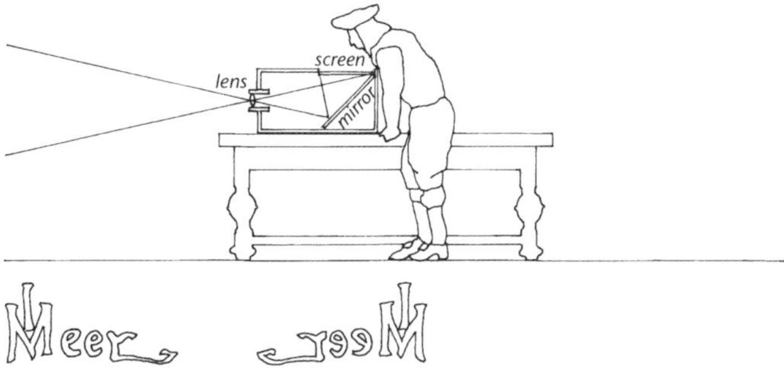

Figure 3 Diagram of a box-type camera obscura with a 45 degree plane mirror and translucent viewing screen. The viewer sees an image that is the correct way up, but reversed left-to-right.

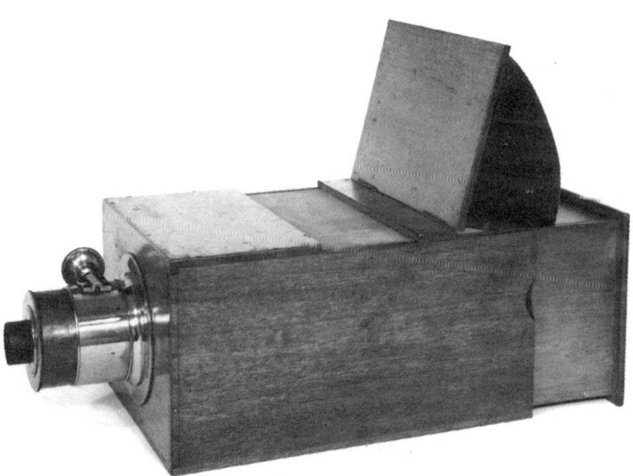

Figure 4
Camera obscura of about 1850, of the type illustrated diagrammatically in Figure 3. The tilted square board and baffles above the viewing screen serve to shield the image from ambient light. This instrument can be focused with a wheel that moves the lens tube in and out.

The invention of the camera obscura was at one time attributed to the Neapolitan 'professor of secrets' Giovanni Battista della Porta, whose compendious *Magia Naturalis* (*Natural Magic*) became tremendously popular in the 16th century and went into more than fifty editions, in several languages besides Latin.[17] It was no doubt the popularity of della Porta's book which gave rise to the misconception. He himself did not make such a claim—although he did imply that he was the first to describe cameras with lenses.

The first edition of *Magia Naturalis* was published in Naples in 1558 and included a description of a pinhole camera, in terms very similar to Leonardo's account. Della Porta greatly expanded his treatment of the camera for a new edition of 1589 to cover the use of lenses and the problem of righting the inverted image. The following extracts are from the 1658 English translation of the second edition:

> Now I will declare what I ever concealed till now, and thought to conceal continually. If you put a small lenticular [convex] Crystal glass to the hole, you shall presently see all things clearer, the countenances of men walking, the colours, Garments, and all things as if you stood hard by; you shall see them with so much pleasure, that those that see it can never enough admire it.
>
> If you cannot draw a picture of a man or any things else, draw it by this means; If you can but onely make the colours. This is an Art worth learning. Let the Sun beat upon the window, and there about the hole, let therebe Pictures of men, that it may light upon them, but not open the hole. Put a piece of white paper against the hole, and you shall so long sit the men by the light, bringing them neer, or setting them further [i.e. adjusting the focus], until the Sun cast a perfect representation upon the Table [i.e. the drawing board] against it; one that is skill'd in painting, must lay on colours where they are in the Table, and shall describe the manner of the countenance; so the Image being removed, the Picture will remain on the Table, and in the superficies it will be seen as an Image in a Glass [i.e. reversed left-to-right].[18]

As for methods for producing a correctly oriented image, della Porta mentions—albeit in slightly vague terms—the use of plane mirrors at angles and an alternative technique involving convex lenses and concave mirrors in combination.

Johannes Kepler learned about the camera obscura some time around 1600, from two sources: from reading della Porta and from working with Tycho Brahe, who had been using the instrument for making solar observations and measurements of the moon. Kepler discusses cameras in both of his works on optics, *Ad Vitellionem Paralipomena* (*Supplements to Witelo*) of 1604, and the *Dioptrice* (*Dioptrics*, i.e. the properties of lenses) of 1611.[19] He himself found astronomical applications for the camera, as well as other uses, it seems, in surveying.

Kepler introduced a method for correcting the image in a camera which, rather than involving mirrors, instead used *two* convex lenses in combination, spaced a suitable distance apart. The principle is shown in the engraving of Figure 5. This comes from the astronomer Christopher Scheiner's *Rosa Ursina*

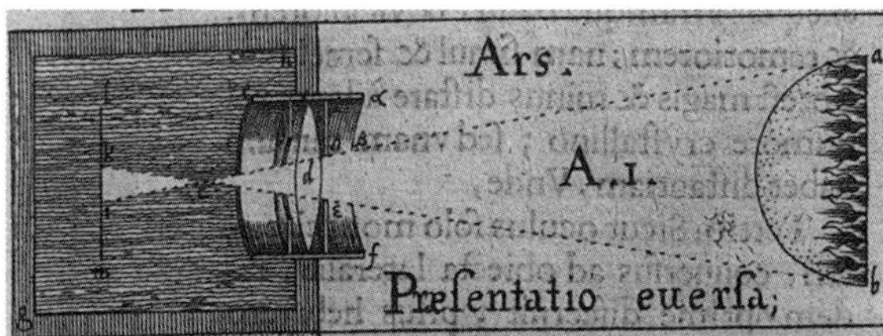

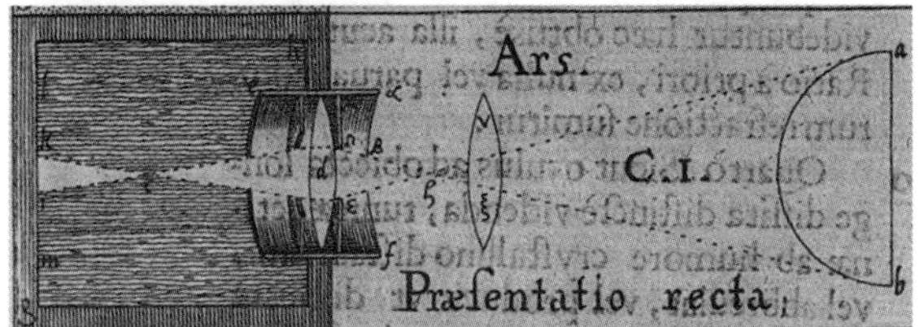

Figure 5
Diagrams of two types of camera obscura; from Scheiner, *Rosa Ursina sive Sol* 1630. The upper diagram shows a camera with a single lens, in which the image is upside-down. The lower diagram shows a camera with two convex lenses, in which the image is now the right way up. (This is just a part of Scheiner's figure: the complete diagram shows other lens configurations and comparisons with the human eye.)

sive Sol (*The Bear's Rose, otherwise the Sun*, 1630) in which Scheiner reports his observations of sunspots made using a combination of telescope with camera obscura—that is, a 'projecting telescope'.[20] The upper diagram shows the single lens camera. The image is upside down ['*praesentatio eversa*']. The lower diagram shows a camera with two convex lenses—Kepler's arrangement—in which the image is now right way up ['*praesentatio recta*'].

The English diplomat and man of letters Henry Wotton met Kepler during a visit to Linz in Austria in 1620.[21] Kepler showed Wotton a portable camera obscura of his own design. It has been suggested that he might have devised this instrument for use in the survey of Upper Austria which he was making at this time.[22] Wotton gives this description in a letter to Francis Bacon:

> He hath a little black tent (of what stuff is not much importing) which he can suddenly set up where he will in a field, and it is convertible (like a Wind-mill) to all quarters at pleasure, capable of not much more than one man, as I conceive, and perhaps at no great ease; exactly close and dark, save at one hole, about an inch and a half in the Diameter, to which he applies a long perspective-trunke, with the convex glass fitted to the said hole, and the concave taken out at the other end, which extendeth to about the middle of this erected Tent, through which the visible radiations of all the objects without are intromitted, falling upon a paper, which is accommodated to receive them; and so he traceth them with his Pen in their natural appearance, turning his little Tent round by degrees, till he hath designed the whole aspect of the field: this I have described to your Lordship, because I think there might be good use

> made of it for Chorography [the making of maps and topographical views]: For otherwise, to make landskips by it were illiberal, though surely no Painter can do them so precisely.[23]

It is clear from what Wotton says that the basic optics consisted of a modified telescope—the cylindrical 'perspective trunk'—with the concave lens removed, leaving a single convex lens.[24] It would have been perfectly feasible for Kepler to have set this tube horizontally and placed the drawing table and paper in a vertical position. Wotton's 'windmill' comparison and his reference to the ease with which the device was turned hint at another configuration, however—one which was frequently used later, in other such cameras designed for landscape drawing. The drawing table and paper might have been *horizontal*, and the lens tube placed *vertically* above this, with a plane mirror angled at 45 degrees above the lens to reflect the image of the landscape down the tube. Figure 6 shows the arrangement in diagrammatic form.

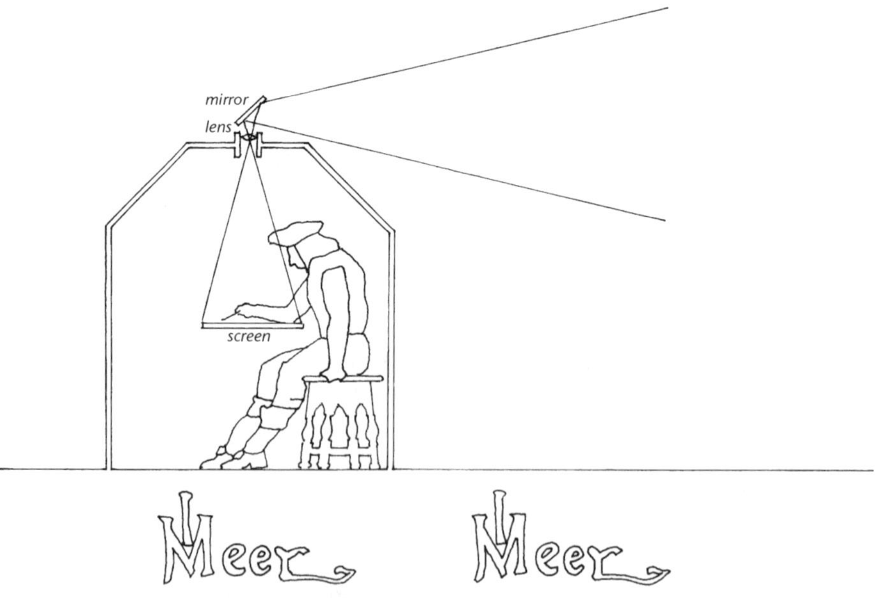

Figure 6 Diagram of a tent-type camera obscura in which the screen is horizontal, the lens is in a vertical tube, and the scene is reflected onto the lens by a 45 degree mirror above the tube. If the artist faces away from the scene, as shown, then the image he sees is correctly oriented—neither upside-down nor reversed left-to-right.

With this arrangement it is sufficient just for the mirror, or a turret housing the mirror, to be rotated (like the cap of a windmill) in the horizontal plane, and the view in any chosen direction to be projected down onto the fixed table. The tent does not have to be turned as a whole. A complete 360 degree panorama can even be traced, given a sufficiently long strip of paper.

Figure 7 reproduces an 18th-century engraving by Giovanni Francesco Costa, in which an artist (at lower left) is using a pyramidal camera of this general type to make, as it seems, an architectural perspective.[25] Note how he ducks his head under the flap of the tent. The turret housing the mirror is at the apex of the pyramid. (It is impossible to tell whether this particular mirror could be rotated independently. The alternative would be for the whole tent to be lifted bodily.) If the user of a camera of this kind faces away from his subject—as Costa's artist appears to do—and directs the mirror backwards over his head, then the resulting image matches the real scene: neither inverted nor reversed left-to-right (see Figure 6).

Whether Kepler's tent-type camera was of this design, it is not possible to

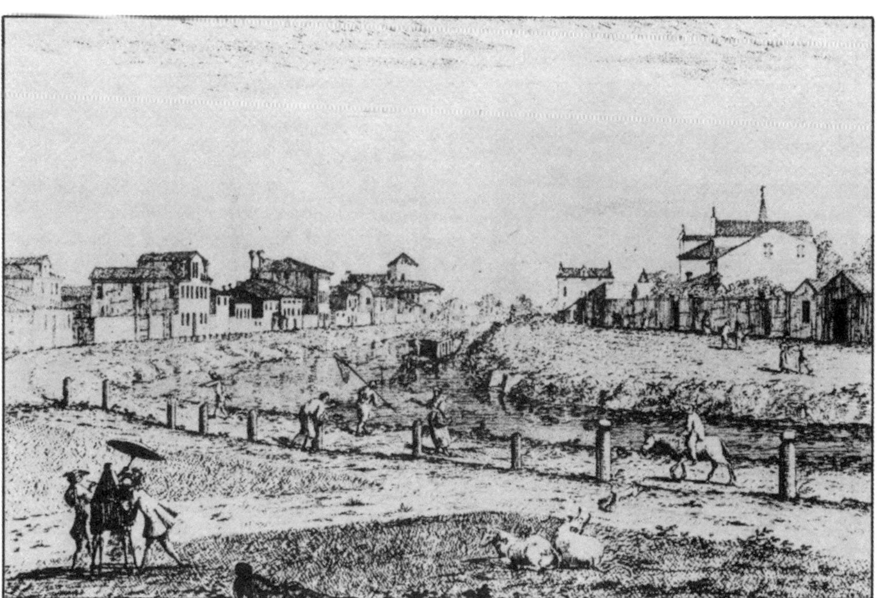

Figure 7 View of the Brenta canal beside the Chiesa della Mira, from Giovanni Francesco Costa, *Delle Delicie del Fiume Brenta* (1750–6). The artist at lower left is using a pyramidal tent-type camera obscura.

know from Wotton's account alone. The arrangement of another portable camera described by the Jesuit scholar Athanasius Kircher is more certain. This is because he includes a drawing of it (Figure 8) in his wide-ranging work on optics, *Ars Magna Lucis et Umbrae* (*The Great Art of Light and Shadow*), published in 1646.[26] Despite the apparent scale given by the human figure, the apparatus was meant to be carried—hence the horizontal poles like those of a sedan chair. J H Hammond, in *The Camera Obscura: A Chronicle*, suggests that perhaps the entire construction was really intended by Kircher to be much smaller than it seems here, and that the artist was meant to fit only his head and shoulders through the square hole in the base.[27] Notice how Kircher's design is for a *double* camera, with lens apertures on two opposite sides. (What might be the utility of this remains a mystery. Perhaps it is just a device of the illustrator, to show one set-up from two different angles.) Each resulting image is projected onto a translucent screen, on which the artist traces, on the opposite side to the lens.

From these examples it is clear that knowledge of the camera obscura was widely spread by the middle of the 17th century, not just among astronomers, but in the popular literature of 'natural magic' and in manuals of perspective drawing methods circulating among architects and painters. We have seen,

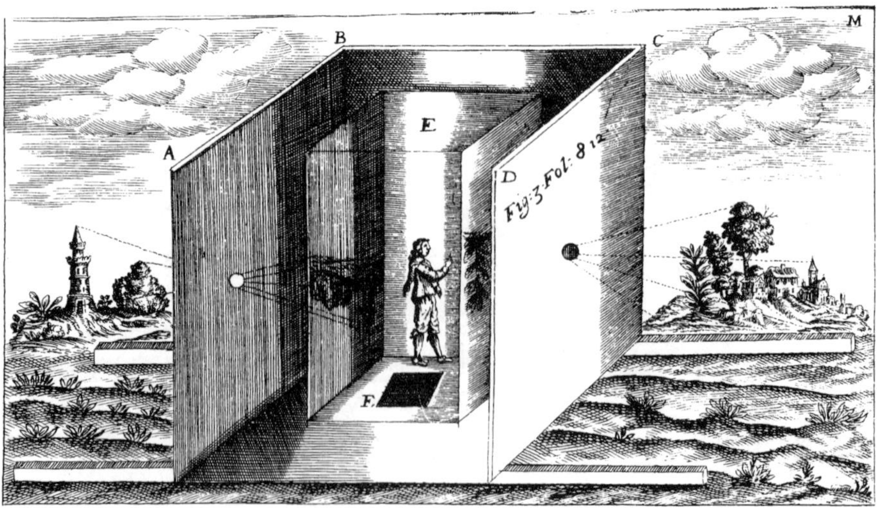

Figure 8 Cubicle-type camera obscura, from Athanasius Kircher, *Ars Magna Lucis et Umbrae* 1646. This actually incorporates *two* cameras facing in opposite directions. In either case, the artist sees the image on the back of a translucent screen. This image is upside-down but not laterally reversed (cf. Figure 2b).

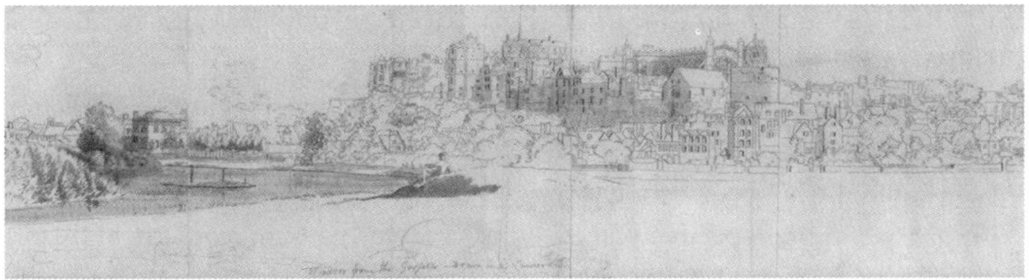

Figure 9 Thomas Sandby, *Windsor Castle from the Gossels*, 1770. The inscription reads 'drawn in a camera T. S.'

from some of the writers quoted, how surprised and delighted people were by their first sight of images formed in a camera. This remains true even of modern viewers of camera obscura images, whom one might expect to have become blasé, with the proliferation of more sophisticated photographic representations. There are, I believe, several reasons for this. Depending on the design of the instrument, the image can be large. The colours can seem richer, more intense, more concentrated than in direct vision—despite the general dimness of the image—particularly when the light comes from behind a translucent viewing screen. The effect is not unlike a large colour transparency seen on a light box. And, curiously, it comes as a shock—given one's preconceptions about photographic prints and slides—to see the objects in the picture *move*. A cloud floats across the sky, the branches of trees sway in the wind, a person innocent of being watched walks across the field of view. It was these last characteristics—the moving image and the sense of spying on an unwary world—which especially appealed to the 16th- and 17th-century commentators. It was the same qualities which turned the camera obscura, in the 19th and early 20th centuries, into a tourist attraction at beauty spots and seaside resorts.[28]

There are no great technical difficulties in making simple outline drawings using the camera. All that is needed is patience, a steady hand, and the ability to make a series of decisions as to where changes of tone and colour should be marked by lines. Figure 9 shows a panoramic drawing of Windsor Castle made by the English 18th-century artist Thomas Sandby with the aid of a camera.[29] As we have seen, in some versions of the apparatus, the image can be projected onto a piece of opaque paper (or board or canvas), and so can be traced directly. In other designs where the viewing screen is translucent and lit from behind, the image must be copied first onto tracing paper or oiled paper. Some means must then be found for transferring the drawing onto a more permanent medium, if required.

Assuming the camera has a simple single lens, and the viewing screen is flat, then the perspective geometry of the image will be exactly the same as that of a conventional linear perspective drawing, or that of a normal photograph. As with photos and perspective pictures there may be certain geometrical distortions produced around the periphery of a camera obscura image, especially where the viewpoint is close to the subject or the angle of view is wide. In an image of a grid of floor tiles, for example, those tiles at the edges of the picture which are nearest to the viewer can begin to seem excessively elongated and unnatural in shape.

It is worth emphasizing certain points about the limitations of the camera and the circumstances in which it can be used, since these have been the subject of misconceptions in the art historical literature. Some writers have imagined that the only designs of camera which artists could have known about and used in the mid-17th century would have been small closed boxes. These could have been of the 'reflex' type with 45 degree mirror and horizontal viewing screen, precisely as in Figure 3; or else of a type without a mirror and with a vertical screen on the rear face of the box, like a photographic plate camera.

Such instruments certainly did exist at this period; however, the first explicit description in print of a box-type camera is in 1657 and the first illustrations appear in the 1670s and 1680s.[30] It was only in the 18th century, it seems, that such cameras were produced commercially and in quantity. In the early 17th century, the written accounts of cameras refer more often to designs consisting of shuttered rooms, booths, or tents.

The overall size of the box camera tends to be limited, for the sake of portability, with a viewing screen that typically might measure 20 or 30 centimetres square. But this is not a fundamental limitation on the apparatus: in booth-type cameras it is perfectly possible to obtain images measuring a metre across or more, quite as big as many 17th-century genre paintings and landscapes.

A second important difference between the box and booth types is that with the box camera the user sees the projected image on the *outside* of the box, under ambient light: it appears as a consequence quite dim. The viewing screen can be shaded to an extent from the sun or other directional light sources, by means of baffles (see Figure 4), but the basic problem remains. By contrast, no light enters a booth-type camera other than through the lens; and the viewer is enclosed *inside* the camera. Once his eyes have adjusted to the darkened surroundings, he sees an image which seems much brighter than that of a box camera.

With box cameras perhaps in mind, some commentators have declared quite firmly that the camera obscura could not have served as an aid to painting or drawing when used indoors, since the image would not have been sufficiently bright. This is simply not the case, as actual experiment shows. The key factors are the level of daylight in the room and the size of the lens and aperture. We will see later how a cubicle-type camera, with a lens of 10 cm diameter, set up in a room illuminated by several tall windows, can give an image which is large, clear, and detailed. Lenses of this size and larger were being manufactured in Holland in the mid-17th century.

There is, admittedly, a tradeoff—as in the photographic camera—between aperture and depth of field. As the aperture is increased to obtain a brighter image, so the depth over which the subject remains in focus is reduced. However, the practical difficulties which this causes to the user have been exaggerated by some sceptical critics. There is not the same need in a camera obscura, as there is with a photographic camera, to have the whole of the image in perfect focus all at one moment. The focus can be adjusted slightly in order to sharpen up different parts of the image corresponding to objects at different depths in the scene. This can be done with relative ease from inside the camera—for example by having the lens mounted in a tube that can slide backward or forward through a circular opening. (It is possible that Kepler's tent camera had its lens in the tubular 'perspective trunk' for just this reason.)

What definite evidence do we have, apart from the publication of local editions of some of the books already mentioned (such as della Porta's *Magia Naturalis*) that artists in Holland in the mid-17th century were aware of the camera and its uses?[31] The letters and memoirs of Constantijn Huygens, secretary to the Prince of Orange and father of the astronomer and mathematician Christiaan, provide one important source of information.[32] Huygens was a lover of literature and art, an enthusiast for the work of contemporary painters (especially Rembrandt, whose greatness he recognized early in the artist's career), but also a keen student of mathematics and the sciences. Huygens especially admired the fabulous Dutch inventor and engineer Cornelis Drebbel.[33] They met in England in 1621 where Drebbel was working under the patronage of James I. Indeed the two men who most impressed Huygens in England were the 'stupendous' Chancellor Sir Francis Bacon and Cornelis Drebbel, 'one unequal in rank but not in talent'.[34] Drebbel had previously studied for a time in Holland with the painter Hendrik Goltzius, whose 'beautiful but disreputable' sister he married.[35] Through Goltzius, Drebbel met Jacques de Gheyn and his son of the same name, both painters and both friends of the Huygens family. It seems probable therefore that Constantijn Huygens came to know Drebbel via the de Gheyns.[36]

Drebbel was celebrated for a whole variety of mechanical and chemical marvels: microscopes, telescopes, magic lanterns, a perpetual motion machine which was said to mimic 'the flux and re-flux of the sea in a glass globe', a 'clavycymbalo' or musical keyboard instrument which played itself through the action of the sun's rays, a kind of air-conditioner which made the Great Hall of Westminster as cold in midsummer as if it were midwinter. The most spectacular of Drebbel's feats was to build a submarine capable of submerging for several hours, apparently with some chemical system for replenishing its air supply. Such a thing would be hard to credit, did we not have Huygens's and other eyewitness accounts of a successful demonstration of the craft in the Thames, with Drebbel at the controls, in front of the King, the court, and several thousand onlookers.

Beside these wonders, the fact that Drebbel also built cameras might seem less impressive to the modern reader, but Huygens was certainly very excited at the prospect of buying one of Drebbel's portable instruments to take back to Holland. On his next visit to London in 1622, accompanied by the young Jacques de Gheyn, Huygens was able to spend much more time with Drebbel. He wrote to his parents:

> The older de Gheyn will be pleased to learn that I will bring the instrument with which he [Drebbel] shows this beautiful picture in sepia, which is certainly one of the master-works of his wizardry.[37]

And when some weeks later he had actually acquired the camera he wrote again, in the most enthusiastic terms, to tell how

> I have in my house Drebbel's other instrument, which certainly produces admirable effects in painting by projection in a dark chamber. It is impossible to describe for you the beauty of it in words: all painting is dead in comparison, for here is life itself, or something more elevated, if only there were words for

> it. Shape, contour and movement come together naturally, in a way that is altogether pleasing.[38]

From the fact that Huygens was able to transport the camera to Holland, we can guess that it took the form of some kind of closed box. He himself described it later as 'an instrument of light construction, so made that when something held outside of it was lighted by a strong sun, it could be projected into a carefully-sealed chamber.'[39] One feature which rather disappointed Huygens, however, and which he felt Drebbel might have improved, was that the image appeared upside-down. It was evidently not therefore the kind of 'reflex camera' illustrated in Figure 3. Perhaps it was more like Kircher's apparatus.

It is worth noting that Henry Wotton was another close friend of the Huygens family. He had been their neighbour when he was ambassador to The Hague in 1615.[40] So here is a further possible route by which knowledge of the camera obscura and its potential for artists—notwithstanding Wotton's misgivings about the 'illiberal' nature of such uses—might have filtered into Holland in the early part of the century.

Back in Holland, Huygens demonstrated the camera to others besides the de Gheyns. He records in his memoirs one particular occasion on which the painter Torrentius (Johan van der Beeck) was present.[41] The career of Torrentius was ridden with scandal and controversy. Accused of witchcraft and crimes against religion, he was imprisoned in 1627 and many of his paintings destroyed.[42] Only a single work of his is known today. It seems from contemporary accounts, however, that his *oeuvre* divided into two quite distinct kinds of picture: on the one hand rather poorly painted nudes and figure compositions, some pornographic or scatological in character, and on the other miraculously realistic and smoothly rendered miniature still lifes.

Huygens was dismissive of the former—Torrentius, he said, was 'disgracefully incapable of painting human beings'—but immensely impressed by the latter. 'I find it difficult', he said, 'to restrain my use of words in asserting that he is, in my opinion, a miracle-worker in the depiction of lifeless objects, and that no-one is likely to equal him in portraying accurately and beautifully glasses, things of pewter, earthenware, and iron so that, through the power of his art, they seem almost transparent, in a way that would have been thought impossible until now.'[43] Figure 10 shows the sole surviving painting: a circular composition of jugs, a glass, and a bridle, with carefully rendered surface

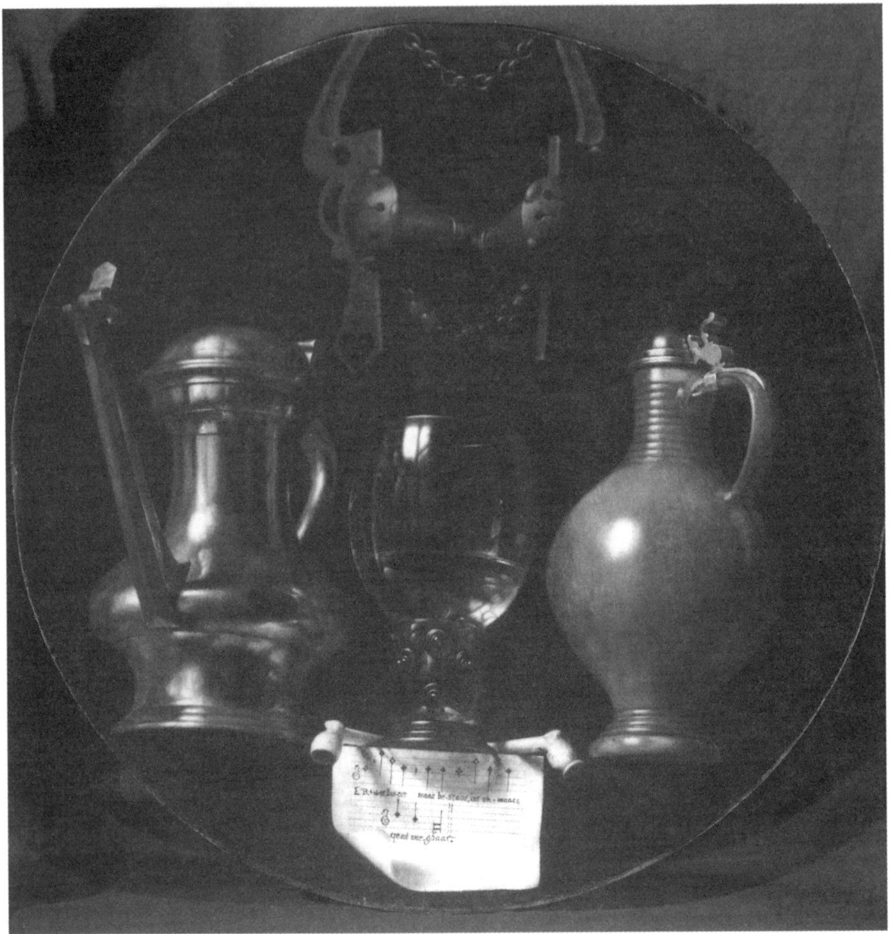

Figure 10 Johannes Torrentius, *Still Life*, 1614, Rijksmuseum, Amsterdam.

textures, highlights, and reflections, once owned by Charles I of England and now in the Rijksmuseum in Amsterdam.[44]

When Huygens showed his camera to Torrentius he got the distinct impression that the painter was feigning ignorance, and that his expressions of amazement were forced and insincere.[45] Huygens discussed this behaviour later with the de Gheyns. They shared Huygens's impression that Torrentius had been trying to conceal his familiarity with the apparatus and that 'the astounding resemblance of Torrentius's pictures to [camera] images' gave strong hints as to the secret of the painter's spectacular technique. (If Huygens was correct in his surmise, then this of course could also account for Torrentius's relative ineptness at figure painting, if for reasons of secrecy, or because of the size of his camera or lens, he had been unable to use it for that purpose. It might also explain why his still lifes were all apparently small in size and, in the case of the Rijksmuseum picture, circular in shape—since the full extent of coverage of a lens in a camera produces a circular image.) Thus a quarter of a century before Vermeer began his career, there were Dutch painters—the de Gheyns—who had certainly seen demonstrations of the camera obscura and at least one painter—Torrentius—whom knowledgeable contemporary witnesses strongly suspected of using the camera in his own studio.

It has been further suggested—though this is wholly speculative—that Vermeer might have developed an interest in optics through a connection with the painter Carel Fabritius, who moved to Delft in about 1650; or, via Fabritius, with his friend Samuel van Hoogstraten of Dordrecht. Both men were fascinated by *trompe-l'œil* and perspective illusion. Most of Fabritius's life-work seems to have been destroyed when the town's powder magazine blew up in 1654, wrecking a large area of Delft and killing the artist himself.[46] But one important 'perspective piece' survives: *View in Delft*, which shows the New Church from the Oude Langendijk. The pronounced distortions of the cobbled street and the seller of musical instruments in the foreground of this picture have provoked critics into a series of conjectures about how it was painted and originally displayed. Among these, Quentin Williams and Peter Kemp have argued that Fabritius used a camera obscura with a curved screen, and that the painting was mounted—or was intended for mounting—on a curved frame, perhaps inside a box with a peephole.[47]

A peepshow box of a different kind made by van Hoogstraten is to be seen today in the National Gallery in London. In this case a series of grossly stretched 'anamorphic' perspectives are painted on the flat inner surfaces of the box. Only when these are viewed from either of two peepholes does the perfect illusion of a series of furnished rooms, complete with dog, fall miraculously into place. At least five more Dutch perspective boxes of this type, all depicting domestic or church interiors, survive from the 17th century.[48]

Peepshow boxes are not cameras, however, and no knowledge of optics is needed for their construction, only of perspective geometry. On the other hand, we know that van Hoogstraten was also interested in the camera obscura;

indeed his writings provide a second source of information about the extent of this knowledge among Dutch painters by the middle of the century. Van Hoogstraten is known to have seen cameras on at least two occasions.[49] And in 1678 he published a book, *Inleyding tot de Hooge Schoole der Schilderkonst* (*Introduction to the Great School of Painting*), in which he discussed the camera and its uses in art.[50]

Some scholars have doubted whether van Hoogstraten could have had much influence on Vermeer in this context, not least because the *Inleyding* appeared after Vermeer's death. They have further argued that van Hoogstraten does not explicitly recommend the camera to artists as a working tool for drawing or composition, citing a passage that seems to see it rather as a visual stimulus or source of inspiration.[51]

> I am certain that the sight of these reflections in the darkness can be very illuminating to the young painter's vision; for besides acquiring knowledge of nature, one also sees here the overall aspect which a truly natural painting should have.[52]

To select just this one remark fails, however, to give a full picture of van Hoogstraten's enthusiasm. For example, he also describes the camera as 'the picture-making invention with which one can paint by means of reflections in a closed and darkened room everything which is outside.'[53] What is more, while it is true that the *Inleyding* appeared too late for Vermeer, this overlooks the fact that the book is devoted in part to reminiscences; and it is clear from what we know of van Hoogstraten's life that he had seen cameras for certain by the 1650s, and very probably by the 1640s. He describes a number of instruments that he has seen on visits at different times in his life to various European cities—'in Vienna at the residence of the Jesuits, in London by the river, and in other places as well.'

> In Vienna I saw countless people walking and turning about on a piece of paper in a small room; and in London I saw hundreds of little barges with passengers and the whole river, landscape and sky on a wall, and everything that was capable of motion was moving.[54]

Van Hoogstraten worked at the Habsburg court in Vienna between 1651 and 1655 for the Emperor Ferdinand III, who greatly admired the artist's *trompe-l'œil* still lifes.[55] At the end of the previous century, Ferdinand's predecessor

Rudolf II had attracted a brilliant circle of painters, mathematicians, and scholars to the Imperial Court in Prague, among whom the most celebrated figures were Tycho Brahe and Johannes Kepler. Ferdinand inherited the treasures of the Rudolfian *Kunstkamer*, and carried on the Habsburg tradition of artistic and scientific patronage. Like Rudolf, he was a keen promoter of astronomy and optics. He was a particular supporter of Jesuit scholars, including Christopher Scheiner, who used the camera obscura for making solar observations, and Gaspar Schott who, like Scheiner, compared the camera with the eye, and was one of the first to reveal the optical image formed on the retina, by painstakingly stripping away the skin from the back of an ox's eyeball. What is more, it was Ferdinand who commissioned the *Ars Magna Lucis et Umbrae* from Athanasius Kircher.[56] There were also scientific instrument makers and optical experts on the Imperial payroll.

This then was the intellectual milieu which van Hoogstraten entered in 1651. It was no doubt his Court connections that gave him an entrée to 'the residence of the Jesuits' and access to their optical instruments. As Celeste Brusati says in her biography of the painter, 'Ferdinand's interests and Van Hoogstraten's concerns clearly met in their mutual fascination with a particular experimental amusement, the image-making device known as the camera obscura.'[57] Brusati suggests indeed that van Hoogstraten's *Feigned Painting of the Imperial Palace Court in Vienna*, dating from 1652, was either painted using a camera, or else was deliberately intended to imitate the appearance of a camera image.[58] The courtyard is packed with crowds of people, horses, and carriages, seen from above as if from an upper window. It is exactly the kind of scene that van Hoogstraten proposes as ideal for viewing in a camera, 'full of countless people walking and turning about.'

Van Hoogstraten returned from Vienna to Dordrecht in the mid-1650s. Whether he was acquainted with Vermeer, and was therefore in a position to pass on his knowledge of the camera, we do not know, but this is at least a plausible possibility. Works by van Hoogstraten were listed in the inventory of Vermeer's possessions made after his death (as also were paintings by Fabritius)—something that strongly suggests a personal connection.[59] It has to be said, however, that there is no other documentary record of any contact. Another option, that Vermeer might have learned about optical instruments not from painter colleagues but from contemporary microscopists, is explored in Chapter 3.

A brief note should be added in conclusion on the *camera lucida*, with which the camera obscura has occasionally been confused. The two instruments are in fact quite different—although they serve similar purposes for draughtsmen and topographical artists. The camera lucida is also quite misleadingly named. It does not consist of a room or box of any kind, but instead is a four-sided glass prism—or sometimes a prism and lens combined—supported on a stand. The artist looks with one eye through the prism at his piece of paper, and simultaneously with the other eye at the object or scene to be drawn, and sees an image of the second superimposed on the first. He can thus trace the image.

The camera lucida is generally taken to be an early 19th-century invention, with the credit shared between the Englishman Wollaston and the Italian Amici.[60] It has however been pointed out by Schwarz that a somewhat similar device was described by Kepler in the *Dioptrice*.[61] Despite this the camera lucida was not generally known about or used by artists in the 17th and 18th centuries. It is improbable that any such instrument would have been available to Vermeer.

2 The discovery of Vermeer's use of the camera

Vermeer was much admired as an artist during his lifetime. He was made headman of the painters' guild in Delft, the Guild of Saint Luke, at the age of only twenty-nine. His pictures commanded high prices. He was apostrophized in verse as being like a phoenix, rising out of the flames of the explosion that killed Carel Fabritius, to take his place as the city's leading painter. This reputation survived after Vermeer's death among the small group of collectors who had acquired his works. But perhaps in part because his *oeuvre* was small and not widely seen—only some three dozen paintings are today attributed to him—his name largely disappeared from view in the 18th and early 19th centuries.

His works were still praised as being exceptionally fine on the rare occasions when they appeared at auction. They were particularly valued in France. Joshua Reynolds was impressed by *The Milkmaid* which he saw on a visit to Holland in 1781.[1] It was because *View of Delft* was so greatly prized by King William and the Dutch state that it was acquired for the Mauritshuis in The Hague in the 1820s—the first Vermeer to pass into a public collection.[2] But otherwise Vermeer tended to be forgotten as a historical personality, and many of his pictures were attributed to more prolific contemporaries such as de Hooch and Metsu. His modern reputation effectively dates from 1866, when the critic Théophile Thoré rekindled enthusiasm with a series of articles in the *Gazette des Beaux Arts* (under the pseudonym William Bürger), for which he had devoted great personal effort and expense to assembling documents, photographs, and a catalogue.[3] Even so, Vermeer's biography and character remained elusive. Thoré/Bürger dubbed him 'The Sphinx of Delft'.

Vermeer's surviving paintings divide naturally by subject into a number of groups, corresponding broadly to successive periods in his career. His earliest pictures, thought to date from between 1654 and 1656, treat Biblical and mythological themes: *Christ in the House of Martha and Mary* and *Diana and her Companions*. From the same period comes the genre subject *The Procuress*, dated 1656 on the painting itself. There are the two architectural views of Delft, *The Little Street*, and the panoramic view of the town, usually dated around 1657/8 and 1660/1 respectively.

There is then the series of small portrait heads and half-length figures of young women, five in number, including the tiny panels *Girl with a Red Hat* and *Young Girl with a Flute*, whose authenticity as Vermeers is sometimes questioned. Interspersed in time with these portraits are the remainder of the paintings, all interiors with one, two, or three figures, which make up the majority of Vermeer's work and which date from the late 1650s up to the very end of the painter's life. It is in these interior scenes and portraits that art historians have most often claimed to detect the use of the camera. At one time or another, however, this suggestion has been made about almost *all* of Vermeer's paintings, the exceptions being the very earliest works painted prior to 1657.

There are no drawings or sketches which can definitely be attributed to Vermeer's hand, and certainly no preparatory perspective layouts of the kind which survive from the studios of other 17th-century Dutch painters.[4] Such documents as have been discovered relating to the painter's life are confined to official records and legal transactions, and none throws light on his painting technique.[5] The inventory of the contents of his studio made for probate purposes after his death lists a few standard items of equipment—easels, palettes, canvases—but nothing to hint at any unusual working method.[6] Certainly there is no mention of a camera or lenses. The house also contained a number of books, but the inventory does not indicate their subject matter. It is not known who was Vermeer's teacher or who studied under him—or indeed if he had pupils at all.[7] Any hypotheses about Vermeer's methods must therefore rest exclusively on one source of evidence: the paintings themselves.

Long before Vermeer's work specifically attracted this kind of speculation, there had been more general suggestions that painters from northern Europe might have employed optical aids. Thus W J 's Gravesande wrote in 1711 how 'It can be noticed regarding the camera obscura, that several Flemish painters (according to what is said about them) have studied and copied, in their paintings, the effects that it produces and the way in which it presents nature; because of this several people have believed that it was capable of giving excellent lessons for the understanding of that light, which is called chiaro-oscuro.'[8] In 1763 Count Francesco Algarotti devoted a very positive chapter of his *Essay on Painting* to 'The use of the camera obscura'. Some of the most skilful Italian landscape painters of the period made use of the camera, according to Algarotti; 'It is probable too, that several of the Tramontane masters, considering their success in expressing the minutest objects, have done the same.'[9]

Several critics and scholars have proposed that the invention and spread of photography may have been not unconnected with the reassessment and revaluation of Vermeer that took place in the second half of the 19th century. The description of Vermeer by the Goncourt brothers in 1861 as 'the only master who has made a living daguerreotype of the red-brick houses of that country' is perhaps just one symptom of this new way of appreciating the painter.[10] Certainly it was the recognition of characteristically 'photographic' effects in his work which triggered the idea that Vermeer himself might have been a camera user. Indeed the first specific suggestion along these lines was made in the *British Journal of Photography* in 1891, by the American lithographer and etcher Joseph Pennell.[11]

Pennell describes how architecturally trained draughtsmen who have no experience of drawing directly from life tend to 'render objects with photographic perspective. So, too, did some of the old Dutchmen. There was a notable example of this in the last exhibition of Old Masters at the Academy, in Ver Meer's *Soldier and Laughing Girl*. But I think it extremely likely that Ver Meer used the camera lucida, if it was invented in his time, for it gives exactly the same photographic scale to objects.'[12] We saw in the last chapter how the camera lucida—which Pennell perhaps confuses here with the camera obscura—although possibly 'invented in Vermeer's time', was almost certainly *not* available to him. This however is by the way. Pennell's essential point is that the perspective of Vermeer's picture is 'photographic'—an effect which would be similar whether produced by the one instrument or the other. The implication of Pennell's comment is that Vermeer might have copied or traced the outlines of an image and in this way obtained relative sizes for the objects depicted.

What he doubtless refers to, more particularly, is the discrepancy in scale between the two figures, of the soldier and the smiling girl (Plate 1). They sit across the corner of a small table, but the officer's head is half as wide again as the girl's. (The disparity seems further exaggerated by the size and blackness of the soldier's hat.) The perspective is perfectly correct in a mathematical sense.[13] It is what would be given by a camera—photographic or otherwise—or by any other kind of apparatus or procedure with which the projected images of objects at the picture plane can be accurately measured and transcribed. The difference in the sizes of the heads occurs because the viewpoint of the picture is very close to the soldier. The effect is possibly less striking to the modern viewer, accustomed to foreground objects appearing very large in snapshots, than it would have been to Vermeer's contemporaries. An artist drawing by eye direct from the subject might have followed more his *psychological* perception of the apparent dimensions of objects, and made the two heads and bodies more equal in size. Certainly this is true of the figures in pictures with very similar subjects and viewpoints by, for example, Vermeer's contemporary and fellow citizen of Delft, Pieter de Hooch (Figure 11).

On the other hand, such 'photographic perspective' cannot in itself be adduced as unequivocal evidence for use of the camera obscura. It could in theory be the result of Vermeer using quite different aids to achieve geometrically correct perspective. For example, he might have viewed the scene through some kind of squared grid of strings or wires, following the principle of Alberti's 'veil' or some of the perspective devices described by Dürer.[14] He might, indeed, just have followed rigorously the geometrical rules of perspective construction.

R H Wilenski in his *Introduction to Dutch Art* of 1928 argued for a technique involving the use by Vermeer of two plane *mirrors*.[15] This idea will be covered in detail in a later chapter. It is worth noting here, however, that one of Wilenski's reasons for believing that Vermeer worked in this way was precisely these same discrepancies of scale in pictures such as *Officer and Laughing Girl*. 'This peculiarity in composition', he says, 'we find nowhere else in European art (except in Hoogstraten) till we get to Degas and the other nineteenth-century artists who exploited similar effects observed in photographs; and it is perhaps one of the ironies of art history that with a Kodak any child might now produce by accident a composition that a great artist like Vermeer had to use all his ingenuity and the aid of mirrors to achieve.'[16]

Lawrence Gowing remarks in similar vein, in a discussion of the perspective of Vermeer's *The Music Lesson*, how very few pictures of its size afford such a wide-angled view. It is this breadth of view 'which gives rise to the apparent

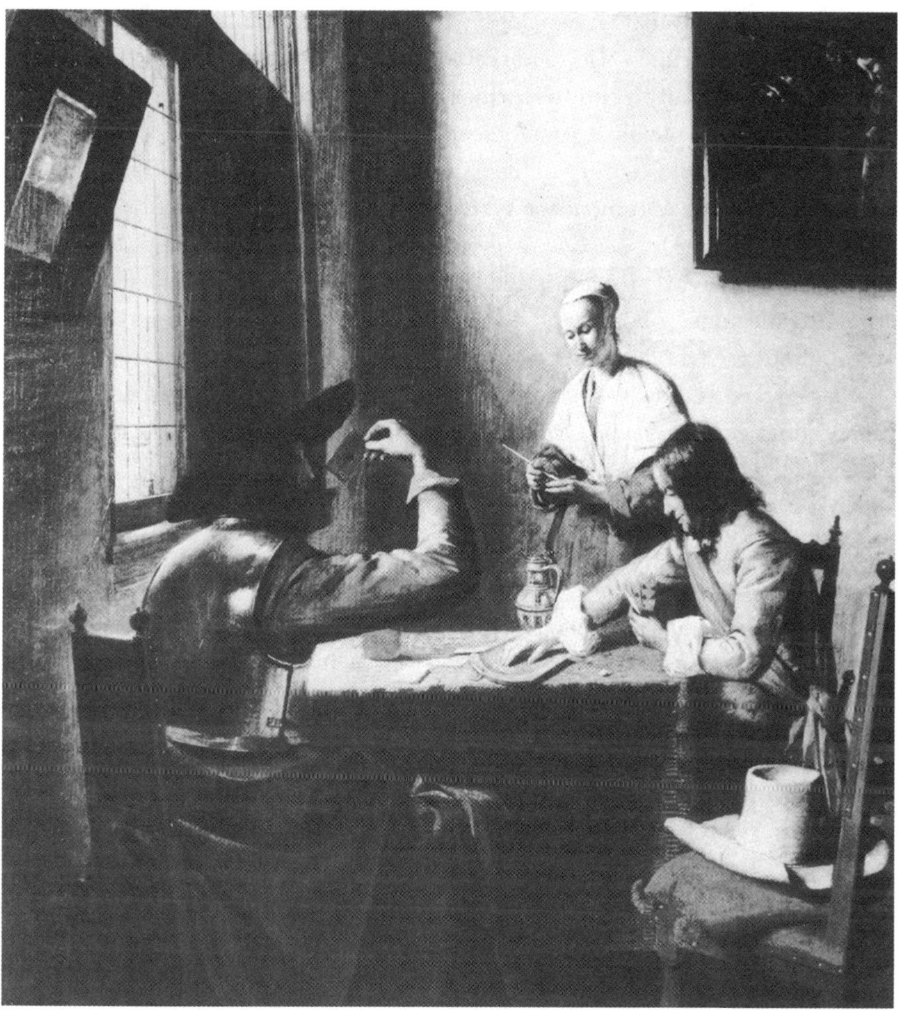

Figure 11 Pieter de Hooch, *Two Soldiers Playing Cards and a Girl Filling a Pipe*, c.1657–8, private collection, Switzerland. Oil on panel, 50.5 x 45.7 cm.

exaggeration of perspective that is often noted in the work of Vermeer . . . It would be a technical task of extreme difficulty to record with conviction so great an angle of vision with no other guide than the naked eye.'[17] Gowing—echoing Pennell and in contrast to Wilenski—attributes this characteristic explicitly to Vermeer's supposed use of the camera. In this connection he links

Vermeer's name to that of Canaletto, who is believed to have used the camera extensively and who is also remarkable for his wide-angle perspectives. Gowing notes incidentally a curious further connection between the two artists: that Canaletto is very likely to have seen Vermeer's *The Music Lesson* when it belonged in the 18th century to his patron, Consul Smith, in Venice.

In the 1940s the argument for Vermeer using the camera obscura was taken some steps further by A. Hyatt Mayor in an article entitled 'The photographic eye'.[18] Hyatt Mayor's subject is the relationship of mechanical means—lenses, drawing machines, photography—to ways of seeing and to style and technique throughout the whole of the history of art. In relation to Vermeer in particular, he has several reasons—besides the pictures' 'photographic perspective'—for suspecting the painter of working from the camera. These have to do with specifically *optical* properties of the camera image:

> In the image cast by a lens Goethe noticed that 'everything is covered with a faint bloom, a kind of smokiness that reminds many painters of lard, and that fastens like a vice on the painter who works from the camera obscura.' Such an effect is one of the fascinations of Vermeer, who used more colors than almost any other great painter, yet blended every combination as perfectly as the ground glass of a camera. Also the high lights on objects in the immediate foreground—the carved lion-head of a chair or the bright threads of a tapestry—break up into dots like globules of halation swimming on ground glass. By throwing near-by objects out of focus, as it were, Vermeer suggested depth with a device more subtle than the standard practice of making them markedly lighter or darker than what is behind.[19]

There are three distinct phenomena described here. First is the blending of colours and blurring of outlines, produced by the light being slightly scattered through the ground glass or oiled paper of the viewing screen. Second is the fact that certain parts of the camera image may be out of focus. Third is the spreading of highlights into 'globules of halation' (see page 32 for a more detailed explanation of this phenomenon). Hyatt Mayor's suggestion is that all three artefacts of the camera—none of them seen by the unaided eye, looking directly at the scene—are actually imitated in the painted surface.

There is without doubt a certain indistinctness to Vermeer's handling of paint in many passages, and a frequent blurring of the boundaries between areas of colour. To point to just a few examples: the backs of the jackets of both the figures in *Woman with a Pearl Necklace* and *Woman Holding a Balance*, or the hair of the figure in *Woman in Blue Reading a Letter*. However, this trait is not any more marked in Vermeer than in some other painters of his period. In some pictures it may even simply be a consequence of abrasion of the surface, or unsatisfactory repainting.

More convincing as the trace of some optical method of working is the undeniable fact that certain areas in Vermeer's pictures, especially in the foreground, are rendered in very soft focus. The most extreme examples, as Hyatt Mayor indicates, are in the skeins of thread which spill from the sewing cushion in *The Lacemaker* and the sculptured brass lions' heads decorating the backs of chairs, which appear in several paintings. But the same phenomenon can be made out in other features too. For example, as Daniel Fink has pointed out, the far wall in *The Milkmaid* is rendered so sharply as to show every nail hole and imperfection in the plaster, while the basket of bread at the front is by contrast rather loosely and fuzzily painted.[20] Again, the map of the Low Countries in *Allegory of Painting* is very crisply rendered, to show not only the printed detail of the map itself and its borders of townscape vignettes, but even the cracks and warping of the pasted paper surface—while the chandelier nearer the viewer is treated in a much more schematic and fluid manner.

Charles Seymour Jr investigated these questions of lack of focus especially in highlights in an experimental study made using an actual antique camera. He published his findings in a paper in 1964 with the nicest title of any on this subject: 'Dark Chamber and Light-filled Room'.[21] Seymour did the work at the National Gallery in Washington, in collaboration with the photographer Henry Beville. They focused their attention on the two small portraits, *Girl with a Red Hat* (Plate 2) and *Young Girl with a Flute*, which are both in Washington. The lions' head chairs feature in both pictures.

Seymour's interest was drawn by the 'dots of heavily loaded pigment' with which Vermeer shows highlights: the mirror-like reflections of light sources in shiny surfaces such as glass, pottery, polished metal, or varnished wood. In reality a highlight takes the shape of the light source which it reflects—although transformed to an extent where the surface is not flat. Thus highlights reflecting the sun approximate to circles or ellipses; highlights which reflect window-panes are rectangles; where such highlights are formed on curved surfaces, the rectangles are distorted into quadrilaterals with curved edges. The highlights on objects in Vermeer's interiors are unlikely to be solar reflections, since it seems probable he would have been painting by north light and they are thus likely to be images of the windows. Nevertheless Vermeer renders these as circles or near-circles.

Seymour proposes that what is happening here is that the images of these highlights have been turned into circles on Vermeer's viewing screen, through deficiencies of the camera lens. Vermeer is imitating these artefacts. If a small highlight of this type, whatever its shape, is not brought exactly into focus at the viewing plane, its image becomes spread out into a 'circle [or disc] of confusion'. Thus what in reality was a rectangular or quadrilateral highlight reflecting a window-pane could appear on the camera screen as a circular blob of light. This happens either if the viewing screen is either too close to the lens or too far away. The 'halation' to which Hyatt Mayor referred is a further diffusion of light that can occur around highlights, as well as around other areas of brightness in a projected camera image.

'Circles of confusion' are therefore just special cases of the general phenomenon of loss of focus in parts of the image. Seymour argues that if this is indeed what Vermeer is doing, then certain conclusions follow:

> the apparatus which formed the image was archaic, for otherwise the points of light in the middle plane would have been resolved and we would not have the record of the discs of confusion in the painting. Either the apparatus could have been adjusted, but for some reason difficult to imagine was not adjusted, or else it was incapable of adjustment.[22]

Seymour believes, from examining a number of early 19th-century camera obscuras, that the latter is the true explanation. He found these cameras to be inherently inflexible and difficult to focus with precision. Attempts to increase depth of field by stopping down the aperture resulted in the projected image becoming unacceptably dim. If this was the case with 19th-century cameras, it would surely have been true *a fortiori* for older instruments. 'In other words, the discs of confusion that occur in Vermeer's painting that would result from poor depth of field in a seventeenth-century apparatus are more to be expected than not . . .'[23]

Seymour and Beville decided to see if they could reproduce these foibles of Vermeer's technique using a 19th-century camera with uncorrected lenses, borrowed from the Smithsonian Institution. They concentrated on the lion's head finial, the background tapestry, and the velvet of the costume of *Girl with a Red Hat*—equivalent examples of which they obtained from museum collections.

Looking at all of these through the camera they found soft-focus effects, and 'circles of confusion' on the lion's head, following Vermeer's rendering closely in their positions and shapes. It did not prove possible to make photographs using the antique camera itself. Instead a modern camera, suitably unfocused, was used to simulate the results, and to produce among others the photo reproduced in Figure 12.

Seymour points out that some of the same soft-focus phenomena are to be detected—although they are less prominent—in the girl's face, especially in the shiny gloss of her lower lip, but that the image comes to a sharp focus at the rear of the head, as can be seen by the definite outline of the white collar. Notice also how the highlight on the pearl earring at the left begins to approximate a quadrilateral. There is a gradient, that is to say, in the degree of focus, from the foreground towards this focal plane.

> The brush is not working at random; rather . . . it adheres with extraordinary fidelity to the system which the photographic reconstruction of the spatial situation called for. The areas are not determined by linear outlines but by light and shade juxtaposed in an orderly way. The order is not that of direct vision; it is that prescribed by an image projected through a lens.[24]

As to the type of apparatus Vermeer might have used for making these pictures, Seymour suggests a small box-type camera of the sort illustrated in Figure 3. Both panels, as remarked earlier, are extremely small. They are of the same order of size as the typical viewing screens of many such cameras: thus the paintings' dimensions might have resulted from Vermeer copying or tracing the projected images directly from an instrument of this kind.[25] In both pictures the viewpoints are very close to the subjects. In the case of *Young Girl with a Flute* the viewpoint appears to be at about the height at which the lens of a box camera would be, if the instrument were set on a table; indeed, the edge of a table is just visible at the bottom of the painting.

Seymour speculates that, contrary to general opinion about their dating, these little portrait heads were among Vermeer's earliest experiments in working with camera images, perhaps from around 1657; and that he developed his technique from this point into the larger-scale interiors with figures which follow. Whether or not he is right about the dates of the portraits, his more general point stands: there is a marked change around this time, not only in Vermeer's subject matter but in his technique. The paintings before 1657 not

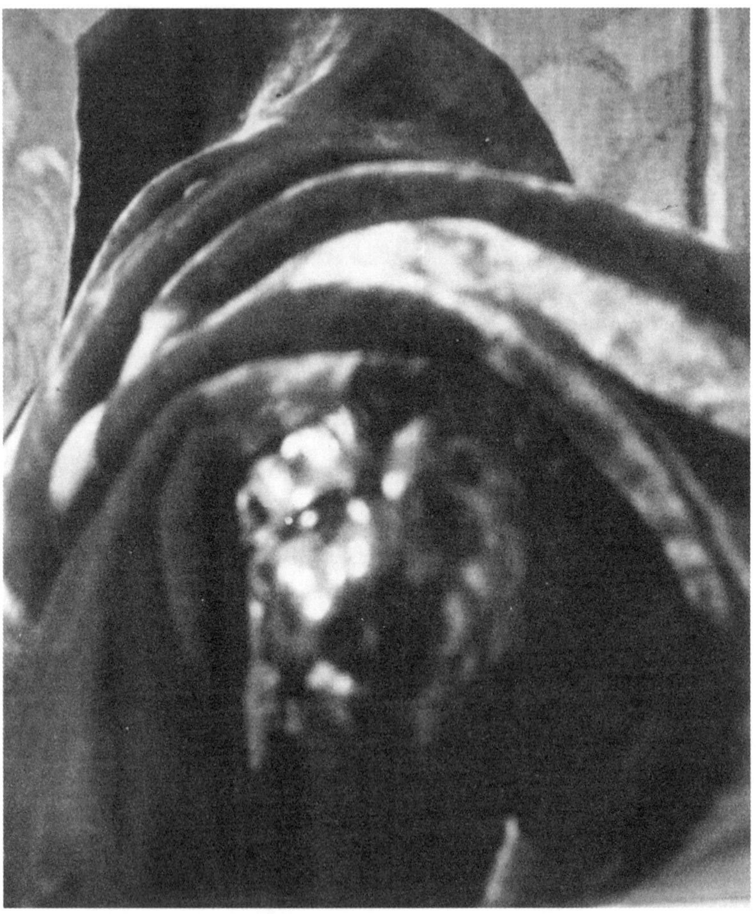

Figure 12 Photograph by Henry Beville to simulate out-of-focus effects in *Girl with a Red Hat* (compare Plate 2).

only lack any of the circular highlights and loss-of-focus effects of the later works, but also, by comparison with the later pictures, their command of perspective geometry is decidedly insecure. This is especially true of *The Procuress*, which seems to have two incompatible viewpoints, one looking up to the figures, the other looking down on a table. This makes a jug set on the table seem very precariously balanced.

One other painting which Seymour discusses in relation to 'circles of confusion' and the camera obscura is *View of Delft* (Plate 3). This he suggests might have been painted with the help of a different size of instrument, producing a much larger projected image—perhaps something more like the apparatus illustrated by Athanasius Kircher (see Figure 8) or a camera in the original sense, formed from an entire room. The painting is large compared with most of Vermeer's other works, measuring approximately 1 m square. P T A Swillens identified, in Willem Blaeu's 1649 pictorial plan of Delft, the actual house from whose windows—as he argued—Vermeer obtained his vantage point (Figure 13).[26] Consistent with his general views about Vermeer's technique, Swillens himself did *not* think that the artist used the room in question as a camera obscura. On the other hand Gowing comments that the picture 'attracts attention in connection with Vermeer's optical procedure for [it has] a suggestion of the effect, the only experience of the instrument that is now common, of the camera obscura often to be seen at public resorts.'[27]

Seymour cites as symptoms of a camera technique, as have other writers, the

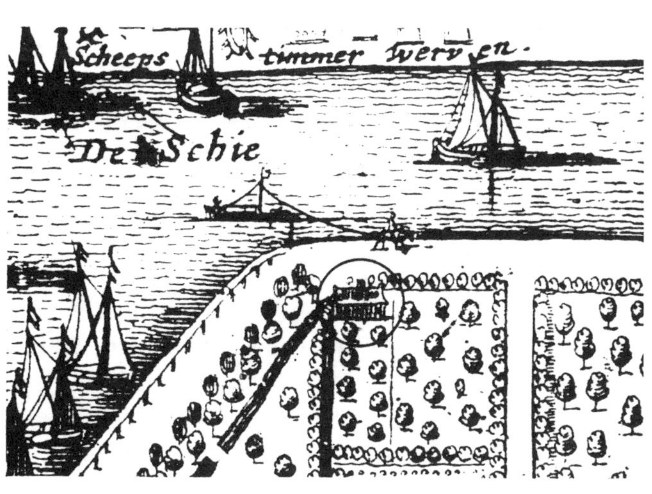

Figure 13
Detail of Willem Blaeu's pictorial plan of Delft of 1648, showing the house (ringed) from whose windows—according to Swillens—Vermeer saw and painted *View of Delft*.

very prominent and numerous 'highlights', represented by circles of white pigment, on the hulls of the two barges at the right of *View of Delft*. This evidence is much less compelling than that of the portrait panels. If these are indeed highlights, they would be produced by the sun, and so would appear genuinely circular to the naked eye—although not as large as they are in the picture. As Arthur Wheelock points out, however, the pattern of illumination of the buildings suggests that this side of the boats is in shadow, not in sunlight.[28]

One should also take caution from the fact that Vermeer uses very similar blobs of white or light-coloured paint to represent the rough textures of some dull and matt materials, including stonework and the leaves of trees here in *View of Delft*, and cloth and loaves of bread in *The Milkmaid*, none of which would produce specular highlights. Wheelock mentions how Aelbert Cuyp used similar small globules of light pigment in his seascapes to capture the 'luminescence of boats and buildings bordering a body of water.'[29] In Gowing's view the *pointillé* in *View of Delft* 'has become disconnected from the substance on which it lies; unhindered by any obvious explanatory purpose it provides a glittering, irrelevant commentary of light.'[30] In the end then it is not any reproduction by Vermeer of artificial optical phenomena that can be taken to indicate that *View of Delft* is a product of the camera obscura. It is rather the uncanny resemblance of the painting to a colour photograph—or better, a projected colour slide. I will take up this comparison again in my final chapter.

The line of experimental work begun by Seymour and Beville was followed up in 1971 by Daniel A Fink.[31] Like them, Fink used an actual box-type camera obscura, in a north-lit room. He observed images on the ground glass screen of his camera, and made tracings by substituting a sheet of clear glass to act as a support, covered with oiled paper on which to draw. He also made photographs. Fink claimed to find evidence of the use of the camera, or of other optical aids, in 27 pictures—that is, almost the whole of Vermeer's *oeuvre* after 1657.

In effect Fink links together the various optical phenomena already discussed. Like Seymour he accepts that, using a 17th-century camera, it would have been impossible to have every part of an image in perfect focus at once—a problem exacerbated by the need to maintain a large lens aperture for maximum brightness. For any given position of the lens with such a camera, there is a plane at a certain depth in the scene at which objects are in sharpest focus. Nearer the camera, or further away, the focus gets progressively poorer. It is in these areas of poor focus that the phenomena of 'circles of confusion' and the halation of highlights would be expected to be more prominent. Fink devotes much of his analysis to this question of quality of focus in relation to the depths of objects in Vermeer's scenes. He also touches on Vermeer's characteristic treatment of mirror reflections—in windows and looking-glasses, not just highlights—and argues that these again, by their very soft focus, betray the marks of the camera obscura.

Unfortunately Fink complicates and weakens his basic case by introducing a number of unsustainable arguments for the use of a camera technique. Some acute criticisms of these ideas of Fink's were made by Arthur Wheelock in his *Perspective, Optics and Delft Artists Around 1650.*[32] For example, Fink cites the fact that the majority of Vermeer's canvases approximate to squares in shape. The complete coverage of a camera lens is circular: of rectangular shapes, it is the square that would cover the greatest extent of this circular image. But of course Vermeer might have preferred near-squares for other reasons. Wheelock points out that square or near-square canvases were common among the work of other 17th-century genre painters.[33]

Fink also focuses on the extensive use of curtains and tapestry hangings in Vermeer's interiors, specifically claiming that their function was to keep the viewing screen in shade, while the main subject was brightly lit. His assumption is thus that Vermeer was using some kind of box camera. If the camera had been an enclosed tent or booth—like Kircher's—then additional drapes would not have been needed for this purpose. There are other plausible explanations however which rule out a direct association with the box camera. It was not unusual in Dutch households for paintings to be covered with protective curtains on runners.[34] Artists came to incorporate *trompe-l'œil* drapes as compositional elements in pictures, in imitation of this habit. For example, there is a famous self-portrait of Gerrit Dou, leaning out of a window and smoking a pipe, with an illusionistic curtain.[35] The green curtain at the right of Vermeer's *Girl Reading a Letter at an Open Window* follows the same convention.

It was also common practice among Dutch painters to use a foreground tapestry or curtain as a *repoussoir*—a compositional device with the purpose of distancing the remainder of the scene. Vermeer is by no means unique in this. As Wheelock points out, a number of such tapestries appearing in Vermeer's paintings are illuminated on their *near* sides, towards the viewer. Thus the part of the room close to the viewpoints of the pictures does *not* seem to be blacked out. Another point worth noting is that interior curtains of this kind were actually used in 17th-century Dutch houses to partition larger rooms and provide smaller private spaces protected from draughts—as shown by some of de Hooch's pictures (Figure 14). So Fink's contention that Vermeer's curtains were there because of the camera is dubious. This is not to deny of course that such draperies, together with the window shutters, play critical roles in controlling the lighting of many of Vermeer's interiors.

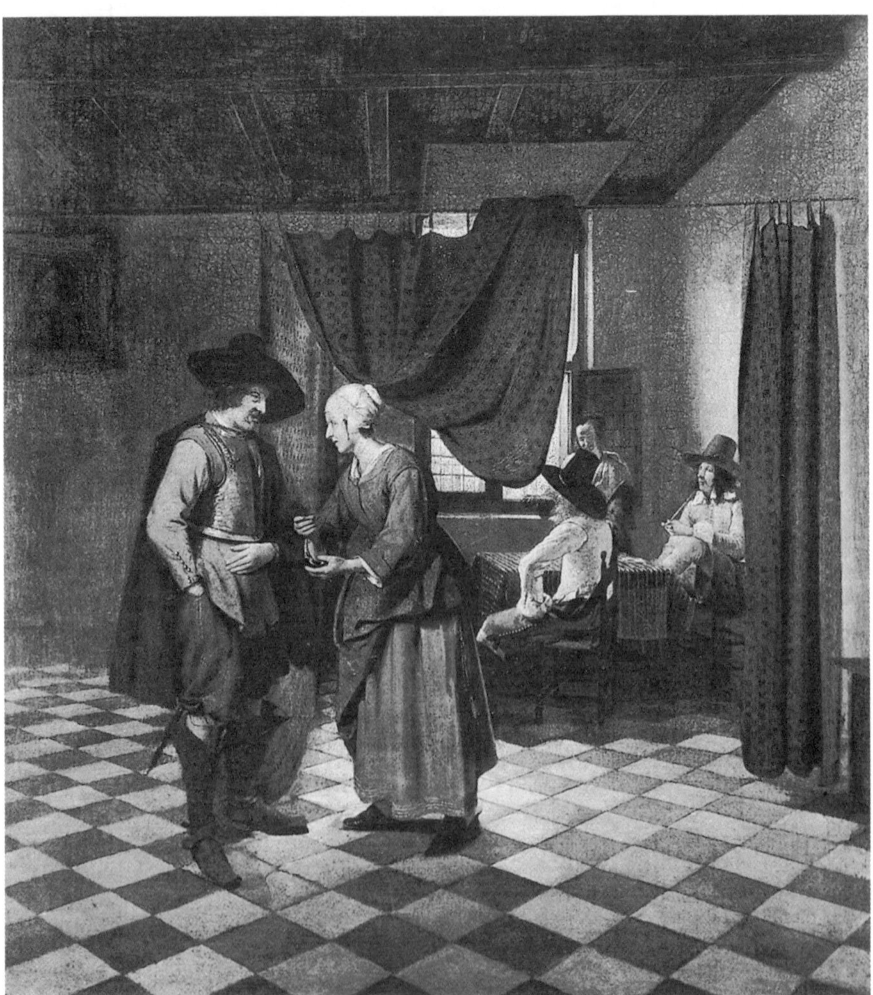

Figure 14 Pieter de Hooch, *Soldier Paying a Hostess*, 1658, Marquess of Bute Collection, Scotland. Oil on canvas, 71 × 63.5 cm.

Finally, Fink mentions as possible evidence of Vermeer using the camera, the fact that he seems to depict the shapes and sizes of objects with considerable precision, though he fails to justify this assertion with any actual measurements. Indeed, the very idea is in some ways logically problematic, since we have no information about the geometry of many of the objects portrayed, other than what can be derived from the paintings themselves. Fink might have strengthened his argument here—curiously he seems to have overlooked the point—by drawing attention to the remarkable accuracy with which Vermeer reproduces the *wall maps* which he includes in several paintings. The historian of cartography James Welu has shown all of these to be actual contemporary printed maps. Examples of every one are preserved in libraries and museum collections.[36] The printed paper sheets were pasted onto canvas mounts for display. (Welu has also identified the various globes, both terrestrial and celestial, in Vermeer's paintings.)

For example, the map of Holland and West Friesland which appears in both *Officer and Laughing Girl* and *Woman in Blue Reading a Letter* was designed by van Berckenrode in 1620 and published by Blaeu some time between 1621 and 1629. Figure 15 reproduces a copy. Here then the real size of the object is known from an independent source, and can be compared with Vermeer's depiction—assuming he represented it with its true dimensions. I will come back to this and other maps in Chapter 5.

It is clear just at a glance that Vermeer's version of the Blaeu map is highly accurate (compare Plate 1 and Figure 15). As Welu himself says, 'If Vermeer did use a camera in *The Soldier and Laughing Girl*, his extremely detailed rendering of the Blaeu–van Berckenrode map suggests that this map served as one of his main points of focus.'[37] Because Vermeer shows this and other maps frontally, square-on, there is no significant perspective distortion involved (other than a reduction in overall size), as there is with objects depicted at angles.[38] The camera obscura was widely used in the 18th and 19th centuries for copying pictures and prints, and for reducing or enlarging their sizes in the process. It is my belief that Vermeer too achieved this indubitable accuracy in reproducing his maps by using the camera, though again there would in principle be other perfectly practicable techniques—for example, the use of superimposed square grids.

Arthur Wheelock's opinion—expressed both in his doctoral thesis and in later monographs on the painter—is that Vermeer most probably *did* make use

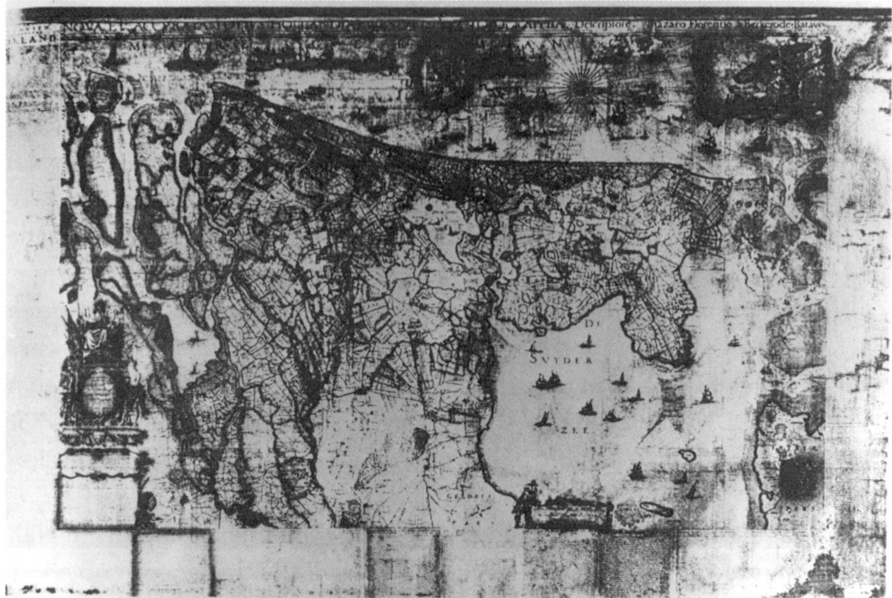

Figure 15 Balthasar van Berckenrode, map of Holland and West Friesland, 1620. Vermeer shows this map in *Officer and Laughing Girl* (compare Plate 1), *Woman in Blue Reading a Letter*, and (very badly lit and in oblique perspective) *The Love Letter*.

of the camera, but in a much more occasional and less systematic way than had been argued by Fink.[39] Wheelock accepts much of the evidence already reviewed here: the passages in soft focus, the photographic perspective, the treatment of highlights (although with certain reservations on this last question, as indicated). In his thesis he describes and illustrates some experiments of his own with a home-made camera obscura which successfully reproduce these phenomena. He says of the two small Washington portraits and *The Lacemaker*, that the visual effects in these pictures are 'so comparable to the unfocused image of a camera obscura that they are almost certainly based on such an image.'[40]

Wheelock's general position, however, is that where Vermeer *did* have recourse to the camera obscura he did not confine himself to copying the image slavishly, but felt free to make many alterations and adjustments. Vermeer was selective, that is to say, and manipulated the optical image to meet higher aesthetic and compositional criteria. Wheelock rejects Fink's proposition that

Vermeer might have traced and copied camera images wholesale and uncritically for more than two dozen of his paintings. In his words:

> Despite the efforts of some art historians to suggest that Vermeer literally transcribed what he saw with optical aids, such was not the case. Although not a single drawing by Vermeer remains, his working procedure must have been much the same as [the architectural painter and perspectivist] Saenredam's: a rough sketch, a construction drawing and then the final composition.[41]

Some of Wheelock's assumptions do, however, need to be questioned. He assumes, for example, that Vermeer would have used a small box-type camera, which would have made it difficult or impossible to see fine detail, or to trace many of his compositions at full size. But we saw in the last chapter how designs of camera were known and used in the 17th century which produced much larger images—as large as the actual sizes of any of Vermeer's canvases. Again, Wheelock argues that the need constantly to refocus a camera obscura to different planes of focus means that it would not have been 'adequately suited as an aid to genre painting'. Should Vermeer have used a camera when working on *The Music Lesson* for example, Wheelock says 'he would have had to refocus . . . a number of times to trace an image in the manner Fink has described.'[42] This I think exaggerates the difficulties—as one discovers from actual practical experience with booth-type cameras. (These issues, and other arguments against the thesis of Vermeer's extensive use of the camera, are taken up again in Chapter 8.) Wheelock's final sceptical conclusion is that 'the camera obscura does not appear to have played a significant role in the evolution of Vermeer's mature style.'[43] In complete contrast to Wheelock, it was the belief of Lawrence Gowing that the influence of optical images was pervasive throughout the whole of the painter's later career and provides a key to understanding Vermeer's special vision and unprecedented style of painting.

I have left Gowing until last in this historical account, and so put him out of chronological order (his monograph was published in 1952), because I believe his to be the most penetrating of all critical accounts of Vermeer's use of the camera. I will give a brief account here of the more technical reasons, beyond those already covered, which led Gowing to these views. I will leave to the final chapter some of Gowing's larger themes relating to Vermeer's visual language and aesthetic.

Gowing's arguments have to do less with the reproduction of any artefacts of lenses than with what he proposes are the effects on Vermeer's style and technique of his transcription of tonal and colour values directly from the optical image. Despite what has been said so far about the use of the camera to make drawings, it is curiously an *absence of linear outline* in Vermeer's finished work to which Gowing points. Vermeer's treatment lacks the boundary lines, or emphases of edges and transitions, by which other artists seek to define and explain three-dimensional form. As Gowing says, Vermeer's rendering of shadow not only obscures line, it interrupts and denies it. He shows an 'almost solitary indifference to the whole linear convention and its historic function of describing, enclosing, embracing the form it limits . . .'[44]

This absence of line seems to extend, Gowing says, to Vermeer's preparatory work on the canvas. He points for example to some striking features of a radiograph of *Girl with a Pearl Earring* (Figure 16) which was made at the time of the van Meegeren forgeries.[45] The first stage of painting as revealed by the X-ray appears to capture a 'sharply contrasted pattern of light and dark'. Gowing

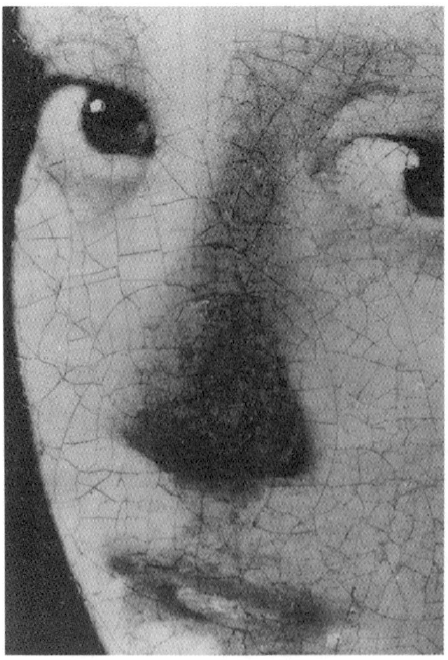

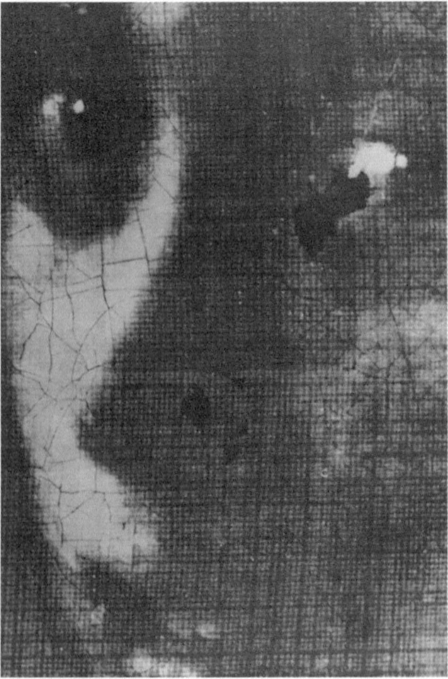

Figure 16 Vermeer, *Girl with a Pearl Earring*, *c.*1665 (detail), Mauritshuis, The Hague. Oil on canvas, 46.5 × 40 cm. X-ray photograph of the same detail.

suggested that this might have been achieved by applying white pigment to a toned canvas. In fact, recent scientific analysis using both X-radiography and examination with the naked eye has shown that it is the ground that is light-coloured (a yellowish white), while the first layer of paint is brownish-black. As the authors of this study explain, 'The purpose of the dark underpaint must have been to brush in a monochrome image on the smooth, light-colored ground.'[46] The underpainted sketch in several pictures is brown, in other cases different dark colours. The paint has been applied, it appears, in solid areas, corresponding to parts of the optical image where the brightness was below a certain threshold. The result is something like that produced in photography by using high-contrast paper, or by the technique of 'posterization'—where the contrast in a print is increased to the point that all intermediate greys are converted either to black or white. In Gowing's words:

> There has been no correction, nor is there evidence of line or design ... No other artist's method reveals this immediate and perfect objectivity: the radiography of painting has indeed never shown a form in itself as wonderful as this strange, impersonal shape. We are in the presence of the real world of light, recording, as it seems, its own objective print.[47]

Gowing is unequivocal in his diagnosis. These curious and exceptional qualities of Vermeer's perception and style of representation have their origin in a reliance on optical images.

> The riddle, or the technical part of it, is plain. And one answer is particularly suggested. It is likely that Vermeer made use of the camera obscura. Certainly it is easy to think that an optical projection was a forming influence on his mature manner. Very possibly the darkness of the instrument's image did not prevent him from using it in the actual execution of many parts of the pictures that we now know.[48]

Gowing goes even further than this. He recalls della Porta's words—quoted in the last chapter—about laying *colours* onto the projected image. Perhaps it was actually possible for Vermeer not just to have set down a tonal basis in monochrome, but to have worked on subsequent stages of the painting process in colour using the camera? We will look deeper into the feasibility of such a technique in Chapter 6.

3 Who taught Vermeer about optics?

The general consensus then among art historians, after a hundred years of speculation and experiment on the subject since Pennell's 1891 article, is that Vermeer seems likely to have used the camera obscura. The remaining debate is about the extent and nature of that use, and the significance—if any—for Vermeer's style. In the rest of this book I present the results of my own researches on these questions. But before I do so, there is one further unresolved historical issue to be explored: from whom might Vermeer have learned about optics and lenses?

It should be said at the outset that this subject is one on which it is possible only to speculate. Unless new documents emerge, the question cannot be answered definitively. My object here is just to demonstrate that there are a few very promising candidates. It has been argued on occasion, against Vermeer's use of the camera, that as a provincial artist he could not have had access to the specialized scientific knowledge and the high-quality lenses that such a technique would have demanded. Nothing could be further from the truth. The possibility was raised in Chapter 1 of Vermeer's fellow painters van Hoogstraten and Fabritius being the sources of information: such links, however, remain hypothetical and undocumented. The alternative is that Vermeer might have been helped by students of optics. Indeed, an active interest in microscopical observation is, curiously enough, one thing shared by several of those rare individuals—outside his family and immediate professional colleagues—whose lives are known to have touched Vermeer's.

The prime candidate for the role of Vermeer's 'optical consultant' is the pioneer of microscopy, Antony van Leeuwenhoek, fellow-citizen of Delft and

Vermeer's exact contemporary.[1] There are absolutely no documents to show that the two men were even acquainted, but there are good reasons for believing in a connection.

Leeuwenhoek is celebrated as one of the first investigators to use the microscope for serious anatomical and physiological research. He did not invent the microscope; indeed he did not even *use* the microscope in the sense in which that term is now understood. The 'compound' microscope with two lenses, an 'object lens' and an eye-piece, was probably invented in Holland—although by whom is disputed—in the first decade of the 17th century.[2] Leeuwenhoek, however, did not use the compound microscope. He made all his observations with single lenses—in effect, with very powerful magnifying glasses.

Leeuwenhoek is famous above all because he was the first man ever to see and describe protozoa and bacteria.[3] That he was able to do this using a single lens is almost incredible. Indeed it was not believed by many historians of science, until Clifford Dobell resuscitated Leeuwenhoek's reputation with his great biography, *Antony van Leeuwenhoek and his 'Little Animals'*, of 1932. It seems that Leeuwenhoek owed his achievements to his very acute eyesight, his skill in grinding lenses,[4] and his endless patience in observation. Nor were his discoveries confined to the 'little animals'. He ranged over many areas of animal and plant physiology, with special interests in the transport of nutrients and in sexual reproduction. He was one of the earliest, although not the first, to study spermatozoa under the lens.[5]

Leeuwenhoek began to win a reputation for his researches in the late 1670s. In 1680 he was made a corresponding member of the Royal Society in London, on the recommendation of his friend the Delft physician Regnier de Graaf, himself a distinguished student of anatomy. Constantijn Huygens, who also knew Leeuwenhoek, supported his case for membership.[6] Leeuwenhoek reported most of his work in a series of letters to the Royal Society. From 1684 onwards some of these letters were collected together and printed.[7]

What are the reasons for imagining that Vermeer and Leeuwenhoek might have known each other? There are two. The first is their close proximity in time and space: they were born in Delft in the same month, October 1632. (Indeed their names appear on the same page of the Baptismal Register of the New Church.) Both lived the greater part of their lives and achieved considerable fame in Delft—a town of only some 25,000 inhabitants—although Leeuwenhoek was away for six years in Amsterdam as a young man, returning around

1654. When Leeuwenhoek married and settled in Delft he bought a house called The Golden Head in the Hippolytusbuurt, just a couple of streets away from Mechelen, the inn on the Market Square owned by Vermeer's family.[8] The Golden Head had been occupied 20 years earlier by the landscape artist Pieter Groenewegen, a friend of Vermeer's father.[9]

The second reason is that after Vermeer died in 1675, Leeuwenhoek acted as his executor. To be more strictly accurate, because Vermeer died owing large sums, his wife Catharina was obliged in effect to declare herself bankrupt and renounce her claims to her husband's estate. The estate was to be disposed of by a trustee or 'curator' nominated by the Aldermen of the city.[10] The appointment of Leeuwenhoek as this curator has been interpreted by scholars in different ways. In 1660 Leeuwenhoek had been made Chamberlain to the Aldermen of Delft. The obligations of the post seem to have been rather light, and the position to an extent an honorific and ceremonial one,[11] but it is possible nevertheless that Leeuwenhoek was asked to organize the disposal of Vermeer's estate as part of his duties as Chamberlain. Obreen, who first published the document in which Leeuwenhoek's appointment is recorded, hints that this might have been a perk of the post.[12] If so, it is hardly likely that the affair brought Leeuwenhoek any profit; it certainly brought him much trouble and effort. The alternative interpretation, of course, is that Leeuwenhoek took on the task as a family friend. Dobell inclines to this idea of a possible family connection:

> To me, the incident appears rather to indicate that Leeuwenhoek may have been a personal friend of the Vermeers, though it also shows clearly that he himself must have held a solid position as a citizen of Delft at that date; since it is inconceivable that the Sheriffs [Aldermen] could have nominated anybody but a respected fellow-townsman to disentangle Vermeer's involved finances. For Vermeer . . . was then rightly regarded as a great artist and ornament of the Town, and his wife apparently had well-to-do connexions.[13]

Wheelock thinks too that the facts suggest a friendship.[14] He searched for records of Leeuwenhoek acting as 'curator' in other cases around this time on behalf of the Aldermen, and could find none.[15] Wheelock also points to another potential link, if more tenuous, between the two men: both were connected with the cloth trade. Leeuwenhoek worked as a draper when he first came to Delft and had been previously employed as a book-keeper in the drapery business during his stay in Amsterdam, while Vermeer's father had had a business weaving the luxurious satin fabric known as *caffa*.

On the other hand, John Michael Montias, who has accumulated a mass of newly discovered documents relating to Vermeer's life and family, is sceptical that the link indicates friendship.[16] He points out that if Catharina or her mother had had any influence on the appointment of an executor, they might have preferred a Catholic like themselves: Leeuwenhoek belonged to the Reformed church. He brings evidence to show that Leeuwenhoek did not act in a way that especially favoured Catharina in disposing of the estate; that indeed he became involved in some dispute with Maria Thins, Vermeer's mother-in-law, about the proceeds from the sale of Vermeer's remaining pictures, especially *Allegory of Painting*.[17] He has discovered recently that Leeuwenhoek *did* act as the curator of other estates besides Vermeer's, on at least three occasions.[18] Finally, and most convincingly, Montias notes the complete absence, despite his own and others' exhaustive researches, of any surviving documented communication between the families, other than in the business of the estate.

What might appear to be another possible line of contact between the two men also turns out to be inconclusive: Leeuwenhoek employed a series of draughtsmen at different times, to prepare illustrations for his letters. These he never refers to by name. Dobell did manage from other sources to identify one Thomas Wilt who worked for Leeuwenhoek in this capacity in the 18th century, and whose father was a pupil of the painter Johannes Verkolje (of whom more below),[19] but no trail leads back to Vermeer.

Although proof of a connection is lacking, suppose just for the sake of speculation that the two men *did* know each other: there remain some suggestive conjunctions and coincidences. Leeuwenhoek returned to Delft to settle in 1654 or thereabouts. It is some three years later, around 1657 or 1658, that the change occurs in Vermeer's style and subjects, from the classical and religious themes of his early career, to the first of the interiors and genre subjects that seem possibly to carry the marks of some optical aid, such as *Officer and Laughing Girl*. It is not clear exactly when Leeuwenhoek might have embarked on his research with lenses, but Boitet says that he started to study various scientific subjects including 'philosophy' (i.e. natural philosophy), astronomy and 'physic' (i.e. medicine) soon after 1655.[20] Later in life Leeuwenhoek himself wrote that he had been making microscope lenses in the late 1650s. (On the other hand Dobell finds no definite evidence of Leeuwenhoek actually using a microscope until 1668, when he went on holiday to England and took an instrument with him.[21])

Leeuwenhoek was certainly an extremely skilful glassblower and lens-grinder. He began to acquire these skills on his own initiative in his youth.[22] It should be said that the type of lens he made for his own microscopes could *not* have served in a camera obscura used as a painting aid. These lenses were tiny near-spheres of glass, with very short focal lengths, held between sheets of brass or silver. (Leeuwenhoek carries a complete instrument in the portrait of Figure 19.) This difference would not, of course, mean that Leeuwenhoek's general knowledge of optics and lens manufacture could not have been useful to Vermeer.

It is more than probable that Leeuwenhoek would have known something about the camera obscura, although at what date it is impossible to say. One of the earliest experimenters with the microscope was Athanasius Kircher whose *Ars Magna Lucis et Umbrae* not only illustrates the camera, but also touches on microscopy.[23] As Kircher writes: 'Who would have believed, had he not used a microscope, that vinegar and milk abound with an innumerable multitude of worms . . . that leaves of plants are made up of tiny filaments, those of Castor Oil Plant contain a collection of star-like bodies . . .'[24] Kircher devoted a whole book to the microscope, his *Scrutinium Pestis* (*An Inquiry into the Plague*) of 1658.[25]

Dobell is scathing about Kircher, dismissing *Scrutinium Pestis* as 'a farrago of nonsensical speculation'.[26] But the value of Kircher's work is not the issue here, simply the question of whether Leeuwenhoek was aware of it. Leeuwenhoek refers to few other workers by name in his letters, and would have needed help studying Kircher's writings in the original, since he did not read Latin. But at least one mention in the letters of 'extraordinarily small living creatures [which] . . . they say, have been seen in Rome' is almost certainly an allusion to Kircher.[27]

Della Porta's *Magia Naturalis* is another book with which Leeuwenhoek could well have been familiar (in the Dutch translation), and which includes descriptions of the camera obscura in the course of an extensive treatment of the properties and uses of lenses more generally.[28] Della Porta's account of the effects of combinations of lenses is rather vague and teasing, but there is no doubt about his reference to their role in magnification, even if he does not clearly describe the compound microscope itself. From the diagrams in Kepler's *Ad Vitellionem Paralipomena* Leeuwenhoek might have found out both about the properties of lenses and the geometry of image formation in the camera (although the book is again in Latin). There are grounds, therefore, for supposing that, had he studied the literature of microscopy and optics in the 1650s, Leeuwenhoek could have discovered much about the camera obscura and its potential uses in art.[29]

Heinrich Schwarz even goes so far as to hint that Leeuwenhoek might have been a camera user himself. He cites a letter to Henry Oldenbourg, secretary of the Royal Society, in which Leeuwenhoek admits to being a poor draughtsman, but says that he has a method of making drawings which he was 'resolved not to let anybody know.'[30] Leeuwenhoek, however, kept his secret.

Finally, there is the vexed but provocative question of whether it is Leeuwenhoek himself who appears in Vermeer's two 'scientific' pictures now known as *The Astronomer* and *The Geographer* (Figure 17 and Plate 4). It was E V Lucas who seems originally to have made this suggestion in his *Vermeer of Delft* of 1922. Lucas's case is somewhat weakened and diffused by his proposition that the microscopist was the subject of no less than four pictures attributed by him to Vermeer, as well as by his acceptance of a portrait by Nicolas Maes, which was at the time exhibited as a portrait of Leeuwenhoek, but is now agreed to have no connection with him.[31]

There *is* one well-authenticated portrait painting of Leeuwenhoek by Johannes Verkolje who worked in Delft from 1673 to 1693 (Figure 18). It is dated 1686.[32] On this basis it shows Leeuwenhoek aged 53 or 54. The microscopist appears seated at a table, wearing a shoulder-length wig and dressed in a golden-brown gown, with a white scarf knotted at his neck. On the table is the vellum scroll, with its red ribbon and seal, which the Royal Society presented to Leeuwenhoek as a 'diploma' of membership. There is also a pair of dividers and a sheet of paper with a drawing of circles and other figures which Dobell identifies as a diagram having to do with scales of magnification. Leeuwenhoek holds a second pair of dividers in his right hand. We have the testimony of no less a person than Constantijn Huygens that the likeness is a good one.

Verkolje also made a mezzotint from the painting (Figure 19), in which the subject is reversed left to right.[33] Now Leeuwenhoek holds one of his single-lens microscopes in his 'left' hand. There is a large conventional magnifying glass on the table, as well as a sprig of oak with oak-apples, marking Leeuwenhoek's

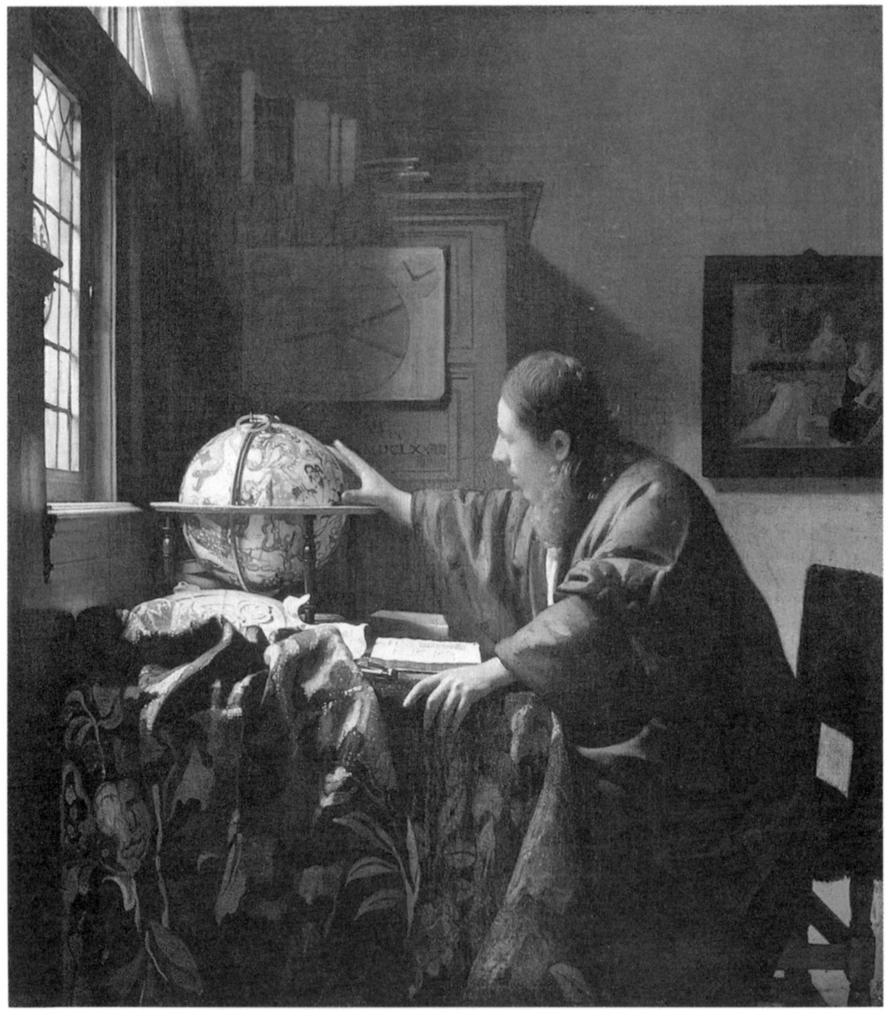

Figure 17 Vermeer, *The Astronomer*, 1668, The Louvre, Paris. Oil on canvas, 50 × 45 cm.

discovery that these galls were produced by tiny insects. Both painting and mezzotint show a globe, apparently a celestial globe of a type that was produced both by Jodocus Hondius and Willem Jansz. Blaeu around 1600.[34]

Is there any resemblance between the man in these Verkolje portraits and *The Astronomer* or *The Geographer*? (Both of these titles are modern, and derive from the items of scientific apparatus that are included in each case.) *The Astronomer*

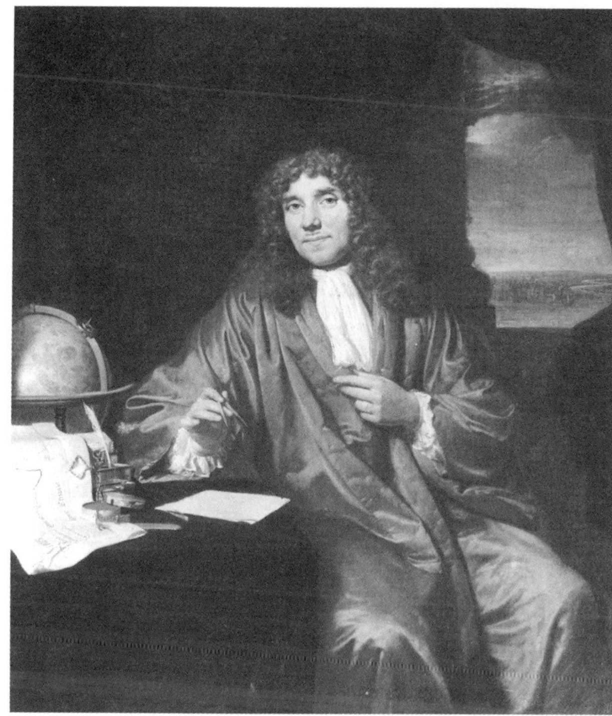

Figure 18
Johannes Verkolje, *Portrait of Antony van Leeuwenhoek*, 1686, Rijksmuseum, Amsterdam. Oil on canvas, 56 × 47.5 cm.

is dated 1668 and *The Geographer* 1669, in both cases on the painting itself.[35] Leeuwenhoek, if it is he, would thus have been in his mid-thirties.

Dobell is not convinced. 'I am entirely at a loss', he says, 'to understand how anyone could seriously suggest that Vermeer's 'Geographers' and 'Astronomers' [that is, including Lucas's two mistakenly attributed pictures] all represent the same person—and that person Leeuwenhoek.'[36] Wheelock, by contrast, is inclined to accept the possibility of Leeuwenhoek being Vermeer's sitter, and remarks on the resemblance between *The Geographer* and the 'broad face and straight, angular nose' of the Verkolje portrait.[37] Montias, however, echoes Dobell's scepticism. He does not, he says 'see any particular resemblance between the elegant, distinguished-looking scholars portrayed in *The Astronomer* and *The Geographer* and the coarse-featured Van Leeuwenhoek.'[38]

I do not myself see Verkolje's Leeuwenhoek as especially 'coarse' in appearance, nor Vermeer's scholars as especially 'distinguished'. Perhaps being disposed to look for affinities, I *do* see something to connect the physiognomies, in

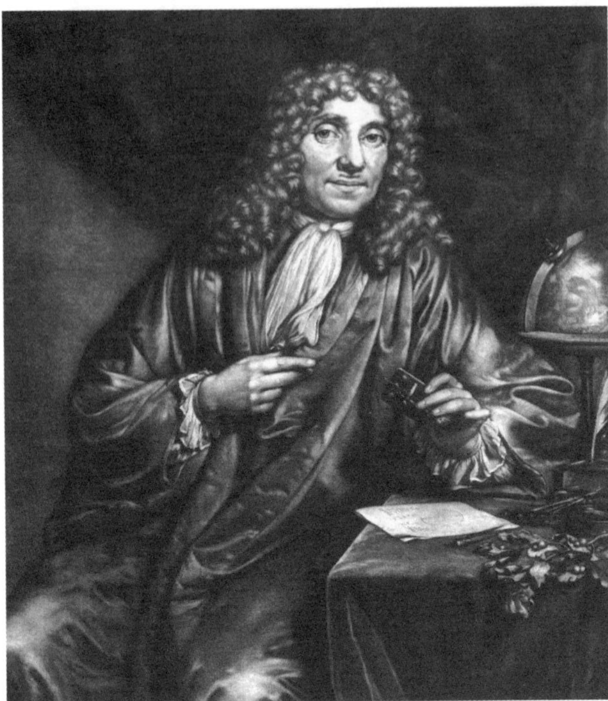

Figure 19
Johannes Verkolje, *Portrait of Antony van Leeuwenhoek*, 1686, mezzotint from the painting of Figure 18. Leeuwenhoek carries one of his microscopes in his 'left' hand. There is a large conventional magnifying glass on the table, next to the sheets of paper.

a certain fleshiness of the features, the long prominent noses, the deep upper lips. What are more difficult to reconcile are the heavy-lidded wide blue eyes of Verkolje's painting with the little button eyes—and are they even blue?—of *The Geographer*.

It was Dobell who saw a possible significance in the *date* of *The Geographer*.[39] In 1669 Leeuwenhoek passed an examination in surveying and was admitted as a *landmeter*. As Dobell says, suppose we give Vermeer's painting the title of *The Surveyor*. Much of the apparatus—the terrestrial globe, the maps, the dividers—seems to fit this label as well as it does *The Geographer*. (Dobell does not mention the fact, but we have seen how Leeuwenhoek had also been studying astronomy in the previous decade.) 'We might then suppose', says Dobell, 'that Vermeer was inspired to paint it by seeing Leeuwenhoek at work on ground-plans and surveys in preparation for his qualifying examination in 1669!'

Perhaps even to speculate along these lines is idle. If the Vermeer canvases *do* show Leeuwenhoek, it is odd that the sitter did not himself acquire either painting. One should also take caution from the fact that both Vermeer's pictures and the Verkolje portrait conform, up to a point, with a longer tradition in Dutch genre painting showing scholars in their studies. There remain some strange coincidences, all the same, between certain items in these particular pictures—setting aside the question of facial resemblances. In all three paintings the sitter wears a flowing gown and a white scarf. 'The astronomer' studies a celestial globe of similar, perhaps identical, design to that which appears in the Verkolje portraits. 'The geographer' carries dividers in his right hand, as in the Verkolje painting. These points gave Dobell pause for thought, despite his doubts. It is certainly pleasant to imagine a congenial friendship between Leeuwenhoek and Vermeer, these two great masters of minute, patient observation of optical images and the effects of light.

It does have to be acknowledged however that any supposed relationship between Vermeer and Leeuwenhoek is based solely on circumstance. On the other hand, there is documentary evidence which indicates that two other distinguished scholars actually met Vermeer: Balthasar de Monconys and Constantijn Huygens. Both have reputations as art-lovers. What has less often been remarked by art historians is that both also moved, like Leeuwenhoek, in the most advanced circles of European optical research. We can only speculate about whether Vermeer was advised by either man: but we have good reason to believe that he came into contact with both.

In Chapter 1 we saw how Constantijn Huygens bought a camera obscura from Cornelis Drebbel in London in 1622 and took it back to Holland. Among his many other interests, Huygens was a keen student of optics and an enthusiastic microscopist: he bought a microscope from Drebbel on the same visit; notes on optics are to be found among his manuscripts;[40] and in his marriage-poem 'The Day's Work' Huygens celebrated his domestic routine with his wife in metaphors drawn from their shared fascination with observations made through the telescope and microscope.[41] Huygens was friendly with Descartes, whom he encouraged to publish his *Dioptrics* of 1637.[42] In the book Descartes illustrates two types of single-lens microscope, the second an arrangement of his own invention.[43] This was later adopted or reinvented by Leeuwenhoek.

Cornelis Drebbel, who sold Huygens the camera, was no mere optical craftsman or technician. He was not the inventor of either the telescope or the microscope, but he was involved in their design and improvement from the very earliest stages. Telescope and compound microscope in their simplest forms are closely related devices, both serving to magnify the object in view, by means of pairs of lenses housed in tubes. The one instrument can in effect be turned into the other by altering the distance between the lenses: hence it is not surprising that the same people were involved in the development of both.

Drebbel was in touch in the early years of the century with the Middelburg spectacle-maker Zacharias Jansen and with James Metius of Alkmaar (Drebbel's own home town), both of them leading claimants to be the inventor of the telescope.[44] Zacharias Jansen and his father Hans also made compound microscopes, and Drebbel had acquired one from them before 1619.[45] Drebbel had microscopes of his own in London by 1622 and one of his instruments reached Rome in that same year.[46] In Rome, Galileo seems to have been making microscopical observations with his new telescope from about 1610. He told a visitor in 1614 that he had

> seen flies which look as big as lambs, and [had] learned that they are covered over with hair and have very pointed nails by means of which they keep themselves up and walk on glass, although hanging feet upwards, by inserting the point of their nails in the pores of the glass.[47]

Galileo's telescope had a convex 'object lens' but a concave eye-piece. The field of view obtained with this configuration, used as a microscope, would have been very restricted. It seems probable that by 1624, when Galileo presented microscopes to several of his friends, he had adopted Drebbel's superior design. This consisted of two convex lenses, in the form of telescope known as 'Keplerian'. According to Clay and Court in their *History of the Microscope*, therefore, Galileo, although sometimes credited with its invention, 'certainly was forestalled by Drebbel so far as the Keplerian form of microscope was concerned, and probably Drebbel was preceded by Jansen.'[48] At other times, it seems, Drebbel made and used single-lens microscopes of high magnification, comparable to Leeuwenhoek's.

Christiaan Huygens, son of Constantijn, was studying geometrical optics in the early 1650s, and began his own treatise on dioptrics at this time, continuing to work on it for the rest of his life.[49] He and his brother Constantijn the younger were building high-powered telescopes in The Hague from 1649. Needing lenses of larger diameter and higher quality than those available commercially, they worked together to grind and polish their own, until Christiaan's departure for Paris in 1655.[50]

Where is the relevance to Vermeer in all this? It has seemed more than likely to some scholars that Constantijn Huygens would have come at some point to know the painter, as his reputation grew, since Huygens was one of the leading authorities on art of his period, and kept in contact with Rubens, van Dyck, and Rembrandt among many other painters. Vermeer's possible connections with Fabritius and van Hoogstraaten, both of them Rembrandt's pupils, have already been noted. The art historian Ben Broos has, however, recently drawn attention to a document that might indicate a closer relationship between Huygens and Vermeer than via these intermediaries.[51]

One Pieter Teding van Berckhout kept a diary in which he recorded a visit to the 'excellent painter' Vermeer on 14 May 1669.[52] He arrived in Delft by barge and met up with 'Monsieurs de Zuylichem [Huygens], van der Horst and Nieuwport'. The last two were political and diplomatic friends of Huygens. Pieter Teding describes how Vermeer showed him 'several curiosities by his hand'. Although he does not say explicitly that the others accompanied him, Broos thinks it highly probable that all four men, including Huygens, were there to see the artist.[53] On a second visit the following month, Teding was shown 'several examples of his art' by Vermeer 'of which the most extraordinary and curious aspect consists in the perspective.' Montias has questioned whether Huygens would actually have gone to Vermeer's studio with Pieter Teding, since it would have been inappropriate for 'such an important Calvinist personage' to have entered the house of a Catholic family.[54] On the other hand, Montias allows that Teding was a close acquaintance of Huygens and he speculates on the possibility that the two friends might instead have met Vermeer in Delft on neutral ground at the house of Pieter van Ruijven, whom Montias believes to have been the artist's chief patron.

Six years previously, in 1663, Vermeer had received another distinguished visitor, the French diplomat and traveller Balthasar de Monconys. Monconys records in the *Journal* of his travels how he went to Delft briefly on 3 August where he admired the tombs of Admiral Tromp, Piet Hein, and William of Orange.[55] Eight days later, on the 11th, he was back again with the sole purpose of visiting Vermeer.[56] The meeting was not, by Monconys's account, a great success. He was disappointed in his hopes of buying a painting. Vermeer had nothing to show him in his studio and took him to see a picture owned by a local baker—probably one Hendrick van Buyten, a wealthy man. Monconys thought the price that the baker had paid was quite excessive for a canvas 'with only a single figure'. The encounter may nevertheless be significant in relation to Vermeer and optics.

Broos notes how, before going to Delft, Monconys had been to see the Huygens family in The Hague, and had passed by again after his meeting with Vermeer. 'One gains the strong impression', says Broos, 'that it was thanks to his contacts in The Hague that the French diplomat had been able to take note of the most famous Dutch artists of that era, such as van Mieris and Dou in Leiden, and Johannes Vermeer in Delft.'[57]

Monconys plays this walk-on part in all biographies of Vermeer, where he is generally introduced as a connoisseur of the arts. This he was indeed: but what fewer writers have remarked on—with the outstanding exception of Heinrich Schwarz[58]—is that Monconys was equally if not more interested in the sciences, and was in touch with most of the leading figures in optics in France, England, Italy, and the Low Countries.

Indeed, while he was with the Huygens family in The Hague, Monconys was comparing their designs of telescopes with his own, and admiring the clarity and sharpness of the images produced by their lenses.[59] On this same trip he went to call on the mathematician Johan Hudde in Amsterdam and on the scholar Isaac Vossius in The Hague to see their microscopes (and dropped into the studio of van Mieris on the way).[60] Both men's instruments had single lenses. Hudde demonstrated to Monconys his methods for melting glass bead lenses and polishing them with salt. He also described his techniques for illuminating specimens. Vossius's microscope consisted of 'a little lens made in the form of a hemisphere, in a wooden frame, which slides behind a little black board, hollowed out on the eye side and pierced at the centre with a tiny hole for looking through'—that is, similar to Leeuwenhoek's.

Like Drebbel, Monconys had himself made significant improvements in the design of the compound microscope. He had introduced a 'field lens', a third lens between the 'objective lens' and the eye-piece, which served to increase the field of view.[61] This innovation has sometimes been attributed to Robert Hooke. But according to Honoratio Fabri in his *Synopsis Optica* (*Review of Optics*) of 1667, the first microscope with a field lens was built to Monconys's specification in Augsburg in 1660, so anticipating Hooke by several years.[62] In 1660 Monconys had also been in Rome, where he had discussed with Athanasius Kircher the latter's microscopical examinations of flies. He was to make a return visit to Kircher in 1664.[63]

Monconys had specially strong contacts with the Royal Society in London, where he met Robert Boyle, the president Lord Brouncker, and the secretary Henry Oldenbourg, among others.[64] With Oldenbourg he went to visit 'Dr Keiffer', Drebbel's son-in-law.[65] He visited the workshop of Richard Reeves in Longacre (or as Monconys has it, 'M. Rives' in 'Longueker'. His spelling of English is highly idiosyncratic. It takes time to work out that when he says he is in 'Oüital' he is in Whitehall). Reeves was an instrument-maker known for his telescopes and microscopes. Monconys was entranced by a magic lantern in Reeves's shop that had a hemispherical crystal lens, and projected little pictures painted on slips of glass.[66]

All this was in 1663, the year of his visit to Holland. Monconys's reception by the Royal Society was by no means an informal one: he was there it seems to represent France semi-officially, on a visit to coincide with a deputation from Holland. As Oldenbourg describes the occasion in a letter to Boyle:

> we had no ordinary meeting; there were no less than foure strangers, two French and two Dutch gentlemen: the French were, Monsieur de Sorbière and Monsieur Monconis; the Dutch, both the Zulichems, Father and Son [Constantijn Huygens the elder, and Christiaan], all foure inquisitive after you.[67]

In summary then: what does all this imply for the questions 'Where might Vermeer have learned about the camera obscura?' and 'Where might he have obtained suitable lenses?' Nothing is certain, but here are two men who visited the painter, Balthasar de Monconys and Constantijn Huygens. Both are in communication with the leading workers in optics, telescopy, and microscopy of the period. Both are sufficiently prestigious in these fields to be received by the Royal Society. Both know Robert Boyle, who mentions portable box-type cameras in his essay *Of the Systematicall or Cosmical Qualities of Things* (1671).[68] Huygens bought a camera from Cornelis Drebbel and is also likely to have learned about Kepler's tent-type camera from Henry Wotton. Monconys knew Athanasius Kircher who illustrated a design of camera. Drebbel, Kircher, Monconys, Huygens, and his two sons, were all either grinding lenses themselves or having optical instruments constructed to their own designs.

At the very least, all this surely disposes of any idea that Vermeer was so remote from the world of scientific optics that he could not possibly have conceived of using the camera; or that, as some authors have claimed, the only type of lens on which he might have laid his hands was a spectacle lens. Chapter 8 will return to these points.

Heinrich Schwarz even goes so far as to suggest that 'the evidence makes it rather possible that one of Monconys's Dutch scientist friends may have called his attention to a painter in Delft who used with some amazing results an optical contrivance and that, therefore, his painting may have had a particular interest and appeal for him.'[69] Monconys however cannot be the sole source of Vermeer's optical knowledge, for he arrives too late. All the changes in the painter's style which can be attributed to the adoption of a camera technique take place around 1657. Monconys makes his visit in 1663, and it is fairly clear from the circumstances that this was his first meeting with the painter, and a brief one at that. If one key individual other than Leeuwenhoek is to be identified, then all the indicators point to Constantijn Huygens as the man who could have helped Vermeer with optics. Huygens has the interest in art; he has the practical knowledge of lenses and the camera obscura; and it is highly plausible, if not definitively established, that he knew the painter in person.

4 A room in Vermeer's house?

All previous theories about Vermeer's use of the camera obscura have been based on studies of the surface properties of the paintings: the evidence of optical effects and the apparent reproduction in the painted treatment of distortions caused by lenses. Critics have also pointed to some general characteristics of Vermeer's perspectives—their unusually wide angles and close viewpoints, and the discrepancies of scale that these produce. In an attempt to provide a different kind of proof of the use of the camera I have approached the subject from a new direction: through an examination of the three-dimensional perspective structures of the pictures. What is the exact geometry of the spaces that Vermeer depicts?

My initial interest was not in the camera at all, but was directed towards answering a more limited question about Vermeer's perspectives. Nearly two dozen paintings, the greater part of his *oeuvre*, show domestic interiors. (They are listed in Appendix A.) Anyone who makes even a passing comparison of these pictures must be struck by the impression that they depict only a small number of different rooms. A few of the spaces, to be sure, are distinctive and unique. Elsewhere, however, certain architectural details recur repeatedly: the same or similar patterns of tiles on the floor, a similar form of timber construction in the ceiling, and, most persuasive of all, some very characteristic patterns of leading in the windows. My first question was, therefore, can we determine from a thorough perspective analysis whether certain groups of paintings do indeed show one and the same room? Is it possible to work out and compare the heights and widths of these rooms, and the shapes and sizes of their various architectural features?

The Dutch art historian P T A Swillens studied the perspective of Vermeer's paintings during the First World War with the same questions in view.[1] (He came to the final conclusion that the perspectives were probably set up *not* using optical aids, but by means of conventional geometrical techniques.) Swillens suggested that some if not all of the interiors showed actual rooms, possibly in one or other of the houses in which Vermeer lived, and which he used as studios.[2] The artist would have rearranged furniture, 'props', lighting, and sitters within the same spaces, for his various compositions. Many readily recognizable musical instruments, maps, carpets, and pieces of furniture (not just the lions' head chairs) also appear in more than one painting. Swillens's earlier analyses have in effect been reworked here, but with the aim of achieving much more rigour and precision.

At the outset I imagined that, if it *did* prove feasible to determine the dimensions of the room or rooms, then this might open the intriguing prospect of identifying their actual locations in the buildings where Vermeer is known to have lived. It has been suggested by several writers, first among them Swillens, that early in his career Vermeer might have had a studio in Mechelen, the inn owned by his family in Delft.[3] The site of Mechelen is known, although the building, sadly, was demolished in the 19th century. It fronted onto the Market Square and backed onto the Voldersgracht canal. A narrow passage, the Old Men's House Alley, ran past its right-hand side to a bridge over the canal behind. Figure 20 identifies the roof of the building in a detail from Dirck van

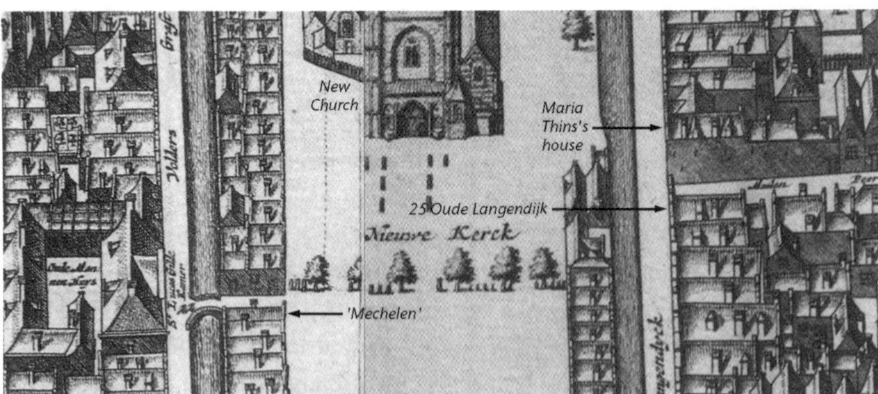

Figure 20 Detail of Dirck van Bleyswijck's pictorial Plan of Delft of 1675–8, showing the site of Mechelen on the Market Square, 25 Oude Langendijk, wrongly identified by Swillens as the house of Maria Thins, and the site of Maria Thins's house, as identified by Montias, on the opposite side of the Molenpoort.

Bleyswijck's pictorial Plan of Delft of 1675–8.[4] The shop that occupies the position today is narrower than Mechelen and the alley has been widened. Although the old building has disappeared, there is an early 18th-century engraving of the Market Square by Leonard Schrenk that records something of its appearance.[5] Figure 21 is an enlarged detail showing the south-facing façade of Mechelen, onto the Square, with the alley visible to the right.

Some time after his marriage in 1653 Vermeer and his wife Catharina Thins moved to a house belonging to her mother, Maria. They were certainly there by 1660.[6] Montias has speculated that Vermeer might have 'kept an atelier in the family inn' after this move,[7] but we know for sure from the inventory of goods mentioned in Chapter 2 that Vermeer had a studio in a 'front room' of his mother-in-law's house at the end of his life. It is clear from the wording of the inventory that this room was on the first floor.[8] Maria Thins's house was on the

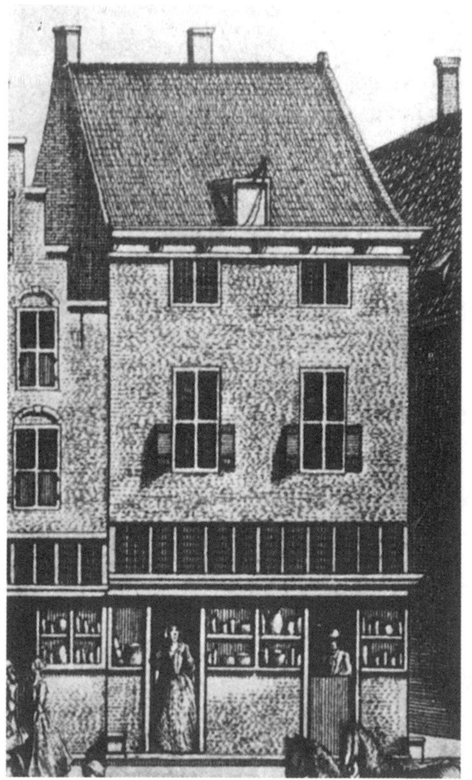

Figure 21
Leonard Schrenk, engraving after drawing by Abraham Rademaker, *View of the Market*, c.1720: detail showing Mechelen. The Old Men's House Alley is on the right.

Oude Langendijk at the corner of the Molenpoort (now Jozefstraat), just across the Market Square from Mechelen. Swillens thought it was No. 25, a house that still stands.[9] However Montias, on the basis of more recent and more extensive archival research, disputes Swillens's identification, and argues that Maria Thins's house was not No. 25, but was on the opposite side of the Molenpoort.[10] This substantial dwelling is clearly visible in the van Bleyswijck map (Figure 20). Vermeer's front-room studio would thus have faced north, looking across the canal towards the New Church. There is no house there now. A neoclassical church was built on the site in 1837, and this was in turn replaced by the present gothic church, Maria van Jessekerk, in 1877. I know of no other pictures of the house besides the van Bleyswijck map and Blaeu's similar *Figurative Map* of 1649.

We will come back to this question of the location of Vermeer's putative studio or studios, armed with dimensions for the room and its windows obtained through the perspective analysis. First let us look more closely and carefully at the architectural details of Vermeer's interiors. As mentioned, a few stand out as being quite distinct. *The Guitar Player* provides the unique case in which a window is shown in a wall at the *right* of the picture; indeed it is the only painting out of 23 interiors in which the light comes from that direction. *A Girl Asleep* is unusual in showing *two* rooms, the second visible through a pair of doorways, in the type of compositional device known as a *doorkijkje*. The far room has a shuttered window facing the viewer. The two women in *The Love Letter* are framed by a doorway close to the picture's viewpoint. Part of a chair, some drapery, and a wall map in heavy shadow are shown in the foreground, on the near side of this opening. This picture is also the only one in which a fireplace appears.[11]

Setting aside these isolated exceptions, there is a large number of other interiors in which the general layout of the room is highly standardized. Every one is a 'central' perspective in which a far wall, without windows, is seen frontally. Often a wall is seen at the left, with one or two windows visible; and even if the windows cannot be seen, the light comes from this direction. In this group of paintings no right-hand wall is ever visible.

The Music Lesson (Figure 22) shares all these characteristics. It is also one of only three pictures in which the *ceiling* is in view. The other two are *Allegory of Painting* and *Allegory of the Faith*, which are sometimes paired in critical discussion although they are works of very different atmosphere, intention, and

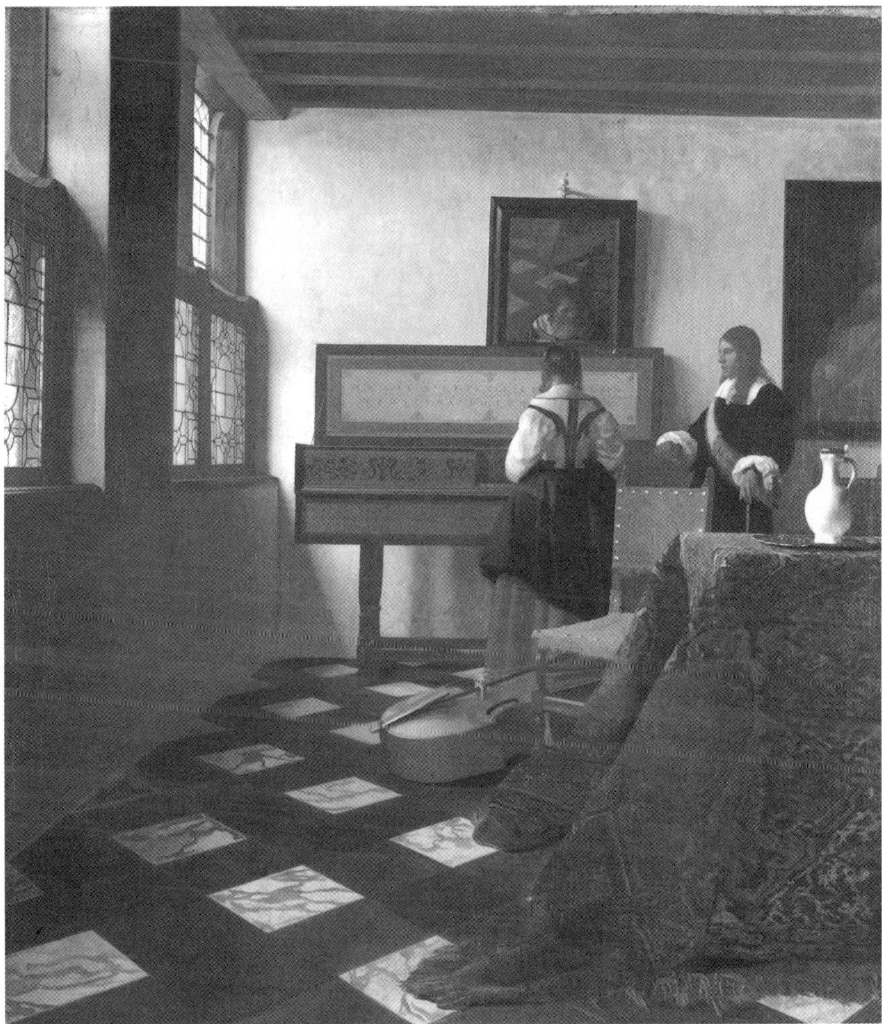

Figure 22 Vermeer, *The Music Lesson*, c.1662–5, Collection of HM The Queen, London. Oil on canvas, 73.3 × 64.5 cm.

quality. They are among Vermeer's largest canvases. The height of the room appears to be about the same in all three cases and the ceiling joists, which run always across the picture, are of similar size and spacing (Figure 23). In *The Music Lesson* it is possible to see that the joists are supported at the left on a timber lintel or wall-plate, running across the heads of the windows.

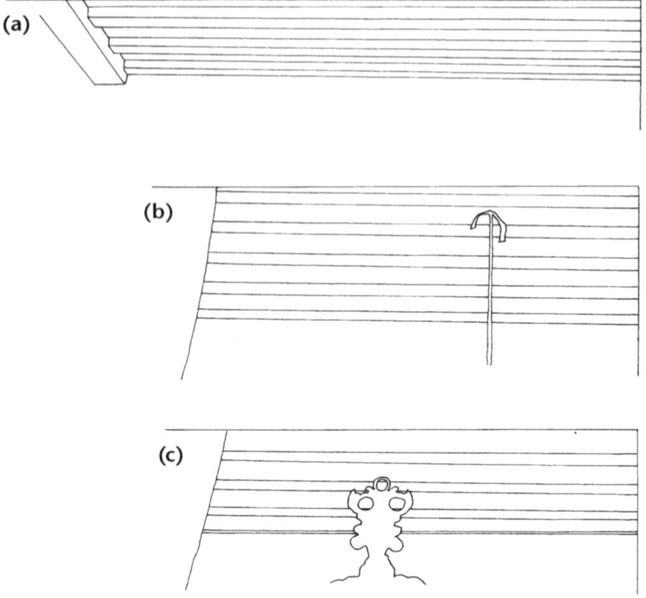

Figure 23
Ceilings visible in three paintings: (a) *The Music Lesson*; (b) *Allegory of the Faith*; (c) *Allegory of Painting*.

The *floor* in *The Music Lesson* is covered in black and white marble tiles (the black with perhaps just a tinge of blue). These are set at 45 degrees to the walls. The veining of the white stone is rendered with a somewhat liquid quality. Closer inspection shows that although seven more paintings show similar marble tiles, the *pattern* in which these are arranged is not always the same (Figure 24). In *The Music Lesson* alone, separate white tiles are framed in a lattice of black stripes. In *Allegory of the Faith* alone, a pattern of what can be read as white Maltese crosses, each made from five tiles, is set on a black background. In the six remaining pictures the colours are reversed, to make a pattern of black crosses on a white ground.[12]

The floor tiles in two paintings, *The Glass of Wine* and *The Girl with a Wine-glass*, are quite different. These are much smaller, ceramic, glazed in greenish-black and pale brown, the colours laid alternately to form a simple checker-board. Again they are set at 45 degrees. It is worth noticing that these two pictures are usually dated among the earliest of Vermeer's interiors, around 1658–60. In five other pictures the room has a skirting made from the familiar blue and white Delftware tiles.[13]

Finally in this inventory of architectural features let us consider the *windows*.

(a)

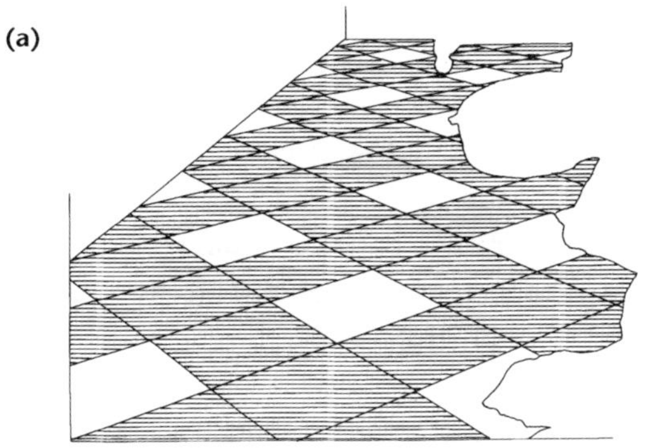

(b)

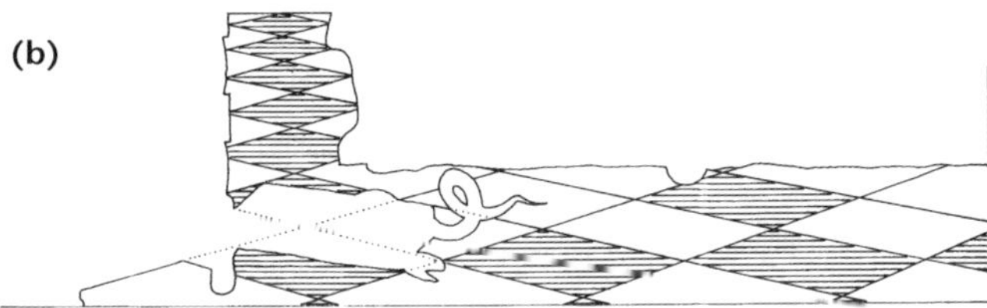

(c)

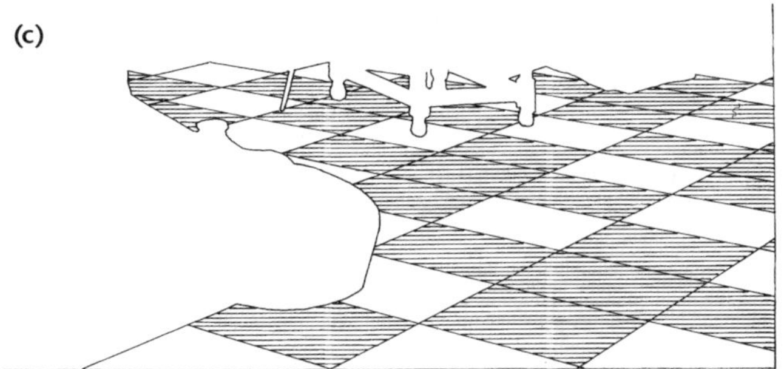

Figure 24
Patterns of marble tiles: (a) the unique pattern in *The Music Lesson*; (b) the unique pattern in *Allegory of the Faith*; (c) the pattern in *The Concert*, which is also found in five other paintings.

In general they conform to a pattern typical of Delft houses of the period. Each is divided into four lights. There are two side-hung casements in the lower half of the frame, and two fixed lights in the upper half. Of all the pictures *The Music Lesson* gives the best view of the arrangement. Such windows were usually equipped with external solid shutters over the lower casements only. The typical design is shown from the outside in Vermeer's *The Little Street* (Figure 25). Vermeer makes repeated use of these shutters, together with curtains, to control the lighting in his interiors. In *Lady Writing a Letter, with her Maid* for example, one of the shutters on the visible window is closed. In three more pictures both shutters on this far window are closed.[14]

Almost all the casements have decorative patterns—some plain, some complicated—formed from the shapes of their leaded panes. Three types can be distinguished. For convenience I have dubbed these the 'lozenge', the 'hourglass', and the 'squares and circles' patterns.

Figure 26 shows all three instances of the 'lozenge' design. This is the type found for example in *Officer and Laughing Girl* (Plate 1). The lozenges occur in the topmost of five rows of rectangular panes. The casements are almost flush with the wall, not set back to form a sill as in the other types of window. There is a short distance from the frame of the window to the far wall of the room. Because of similarities in its overall design and positioning I have included the window in *The Milkmaid* in this group, despite the fact that this seems to lack the defining diamond-shaped panes in the top row.

The 'hourglass' design is peculiar to the pair of matching pictures of scholars in their studies, *The Astronomer* and *The Geographer* (Figure 17 and Plate 4). Here the hourglass shapes form the topmost of *six* rows of rectangular panes (Figure 27). The window frame is a more sophisticated piece of joinery than in the 'lozenge' type. Both casements and upper fixed lights are surrounded by elaborate mouldings. There is a similar distance from the frame to the far wall in both cases. The same large cupboard occupies this position in the two paintings. Indeed the only difference is that the unshuttered casement in *The Astronomer* has a central circular stained-glass pane with a yellow and orange motif.

The most complex pattern of window-panes and leading occurs in at least eight paintings (Figure 28) and perhaps more. Here the entire casement is organized in an arrangement of interlocking squares, circles, and semi-circles, with a central panel of four squares bounded by four semi-circles. In *The Music Lesson* we see two windows of this 'squares and circles' type. Both casements are visible in the farther window, one casement only in the nearer window. We can also see the whole of one of the fixed lights, which has an hourglass motif in its topmost row of panes. The casements themselves are set back to provide shallow window ledges. The fixed lights are recessed even further, with their

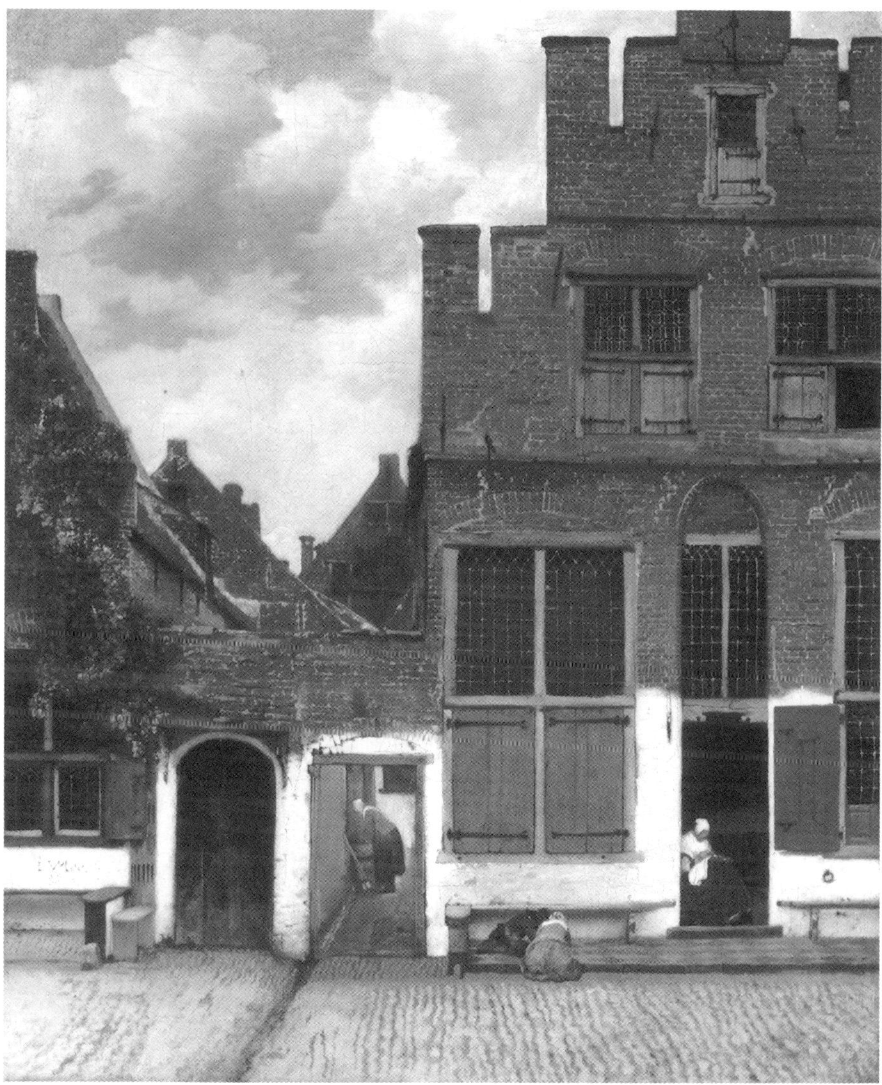

Figure 25 Vermeer, *The Little Street*, *c.*1657–8, Rijksmuseum, Amsterdam. Oil on canvas, 54.3 × 44 cm.

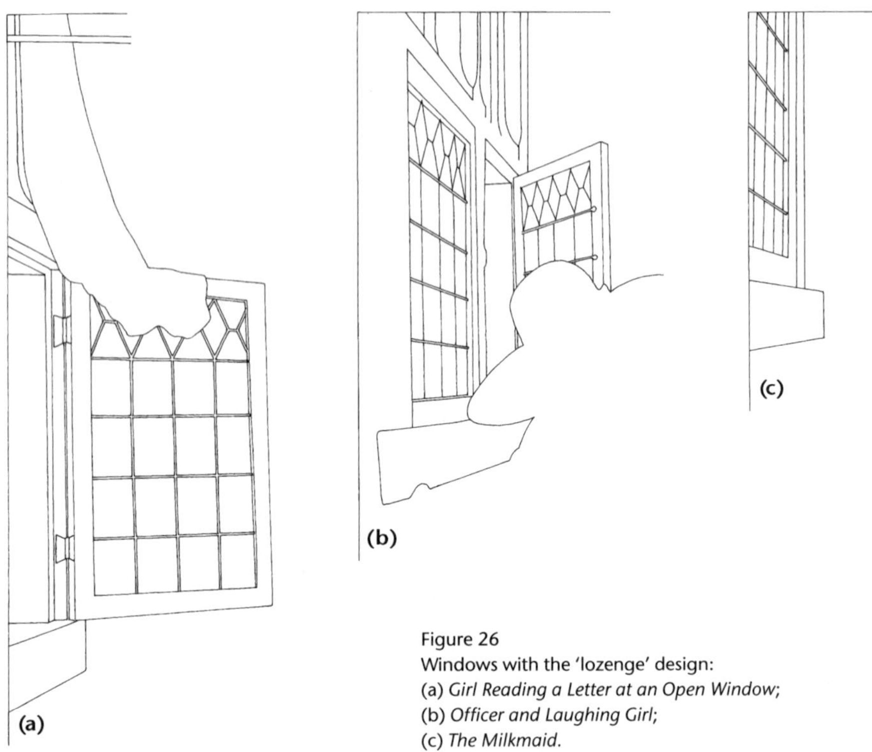

Figure 26
Windows with the 'lozenge' design:
(a) *Girl Reading a Letter at an Open Window*;
(b) *Officer and Laughing Girl*;
(c) *The Milkmaid*.

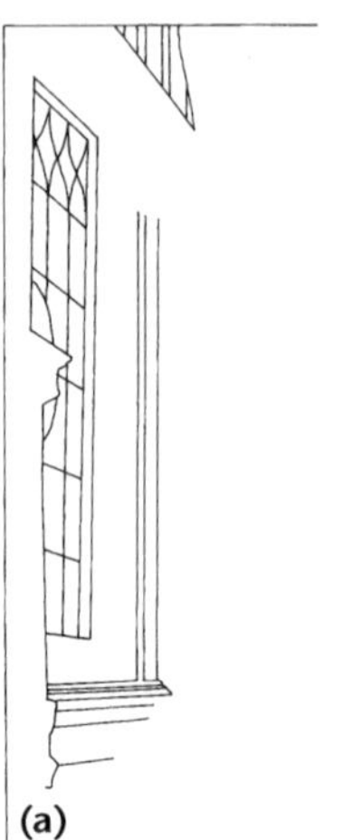

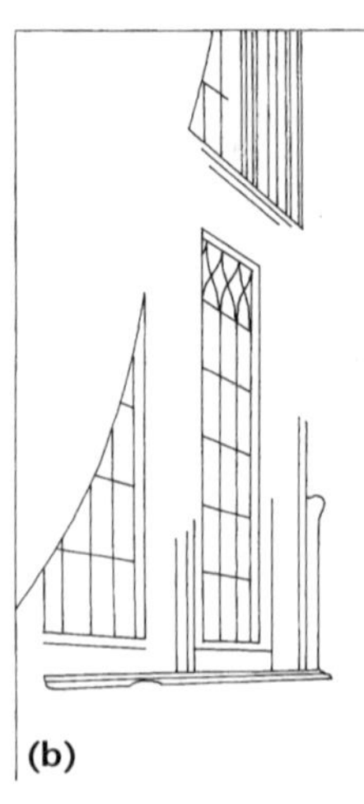

Figure 27
Windows with the 'hourglass' design in:
(a) *The Astronomer*; (b) *The Geographer*.

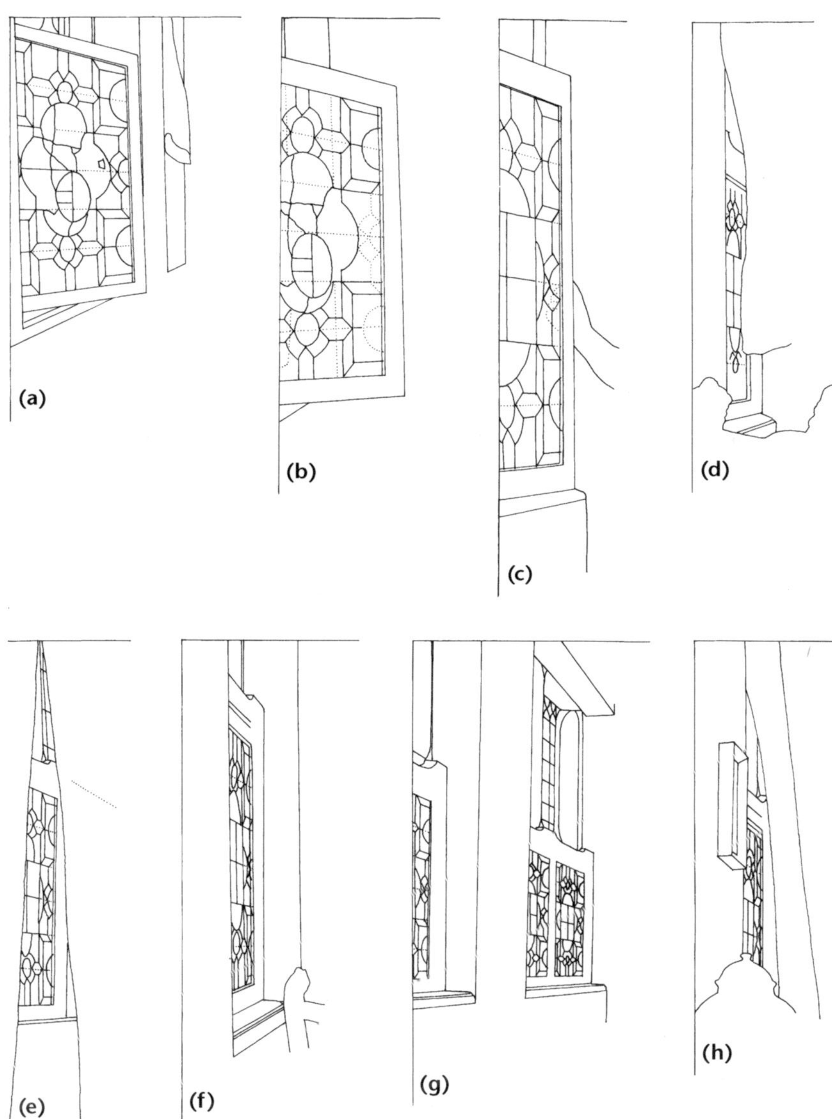

Figure 28 Windows with the 'squares and circles' design: (a) *The Glass of Wine*; (b) *The Girl with a Wineglass*; (c) *Young Woman with a Water Jug*; (d) *Woman with a Lute*; (e) *Lady Writing a Letter, with her Maid*; (f) *Girl Interrupted at her Music*; (g) *The Music Lesson*; (h) *Woman with a Pearl Necklace*.

surrounding frames heavily chamfered (cut away at 45 degrees). Unlike both the 'hourglass' and the 'lozenge' types, these windows—in most instances—end flush with the far wall.

There are two exceptions, *Girl Interrupted at her Music* and *The Girl with a Wineglass*. Here we might expect to see two windows as in *The Music Lesson*, but we see only one, nearer the viewpoint. It is just possible nevertheless that in each of these pictures a second further window *was* originally shown, since the backgrounds of both have been heavily overpainted.[15] Faint traces seem to persist. Had the windows been completely shuttered they would in any case have been in deep shadow.

In most casements of this 'squares and circles' type the glass is plain, perhaps with a blueish tinge. In the windows in two paintings however there is a stained-glass centre-piece, predominantly red and blue, depicting the figure of Temperance and incorporating a coat of arms.[16] These are the same two pictures—*The Glass of Wine* and *The Girl with a Wineglass*—that show the ceramic tiled floors. A second such coat of arms decorates the casement in *Lady Writing a Letter, with her Maid*.

Finally a fourth design of window is to be found uniquely in *Lady Standing at the Virginals*. The pattern in the casements is similar to the 'lozenge' type. But the general arrangement of the sill, the setback of the fixed lights, and the position of the window flush with the far wall, are all equivalent to the 'squares and circles' type—although the mouldings on the upper part of the frame are more refined. Because the shutters are closed, and it is painted in deep shadow, it is impossible to tell whether the window in the companion piece *Lady Seated at the Virginals* is similar or is of the standard 'squares and circles' design.

What then do all these various details of windows, floors, and ceilings seem to suggest about the rooms that the paintings show? Of course, to ask the question in this way risks the accusation of 'naïve realism'—the danger of assuming a priori that these rooms were real ones, and that Vermeer conscientiously made literal transcriptions of their precise appearances. This would be to overlook the possibility that, like Pieter de Hooch, he painted some actual room or rooms repeatedly, showing the same general arrangements and some of the same features, but introduced variants in detail—in the glazing, in the tile patterns—to suit the needs of the various compositions, or even at whim. Beyond this, there is no reason why the artist might not have assembled architectural components from different actual rooms into many combinations; or indeed that *all* these features are perfectly imaginary and never corresponded to any reality at all.

But suppose that we *do* see a series of actual rooms: what consistencies can we find? Can we say something about the number of distinct rooms that might be involved? To come to any such conclusions it is necessary to be systematic, as in the synoptic table in Appendix A, which lists all the various characteristics. My reading of the implications of these data is as follows.

The three unique cases, where we see two rooms, or where the light comes from the right, must be set aside. Apart from these there are some promising pairings or larger groupings. The strong impression given for example by comparing *The Geographer* with *The Astronomer* is that we see the very same room, with the identical furniture only slightly rearranged. The stained-glass insert in the astronomer's casement is the one minor detail that differentiates the two 'hourglass' windows. (On the strength of Chapter 3 we might even be tempted into imagining that we are being granted a privileged glimpse here into Leeuwenhoek's own study.) Certainly Swillens came to this conclusion. He distinguished five rooms in all, of which this was the first (see Appendix A for full details).[17]

There are three paintings, as we saw, in which the 'lozenge' design of casement appears. Going by the windows alone we might imagine that these again show one and the same space. This was Swillens's second room. No other features are visible in these three pictures, however—no ceilings, no floor tiles—by which to test such a hypothesis. A floor is visible in *The Milkmaid*, but this seems to made of wooden boards. The only tiles are in the Delftware skirting. In what follows I will describe ways of making a more quantitative analysis of the geometry of some of the rooms, in order to determine the precise dimensions of windows and furniture. However, this is only possible for those pictures where a reasonably large area of the floor is visible and that floor is *tiled*. This criterion counts out all pictures in which the windows are of the 'lozenge' and 'hourglass' types. With them we can pursue this line of inquiry no further.

So far as the evidence of the windows goes, we are left with the eight paintings with 'squares and circles' casements, plus *Lady Standing at the Virginals* and *Lady Seated at the Virginals*, where the windows are of a similar if not identical design. This group includes *The Glass of Wine* and *The Girl with a Wineglass*, where the floor tiles are of the small ceramic type. Swillens's solution here was to propose a further *three* rooms, the overall dimensions of which he imagined were similar: one for the pair of ladies playing the virginals, a second for the two paintings showing the ceramic-tiled floors, and a third for the remainder. His last room covered, therefore, six instances of the 'squares and circles' windows, all of them combined with marble tiles on the floor. With these, he grouped four more pictures in which no windows can be seen, but again the floor is marble. Swillens thus believed that *ten* paintings showed the very same space. This meant, however, that he was obliged to overlook some of the discrepancies. He argued that in the matter of the varying patterns of marble tiles 'the painter allowed himself some licence.'[18] The perspective analysis which follows will throw more light on this and some of the other anomalies in the window details.

5 Reconstructing the spaces in Vermeer's paintings

There are 11 paintings by Vermeer in which a significant area of tiled floor appears (whether marble or ceramic). In every such case it is possible to reconstruct the three-dimensional geometry of the room, by working backwards from the two-dimensional perspective image embodied in the painting. This was my next step. I made no assumption at the outset that some or all of the various rooms might be the same. My purpose was to test this possibility by the results of the analysis.

The original methods for constructing measured perspective views, formulated in the 15th century by Alberti, Leonardo, Piero della Francesca, and others, were essentially techniques for setting out images of gridded horizontal surfaces—such as, precisely, tiled floors. The painter, knowing the true three-dimensional shape of any object, figure, or building that he wanted to depict, and knowing its position on this ground surface, could derive a perspective image of the object by reference to the grid. I followed a procedure that essentially reverses this process: to infer, from the two-dimensional perspective image, the three-dimensional forms of the room and furniture that the image represents.[1]

Strictly speaking this process of inference cannot be completely deterministic. In principle there are infinitely many three-dimensional forms that could correspond to a given two-dimensional perspective picture. This fact provides the basis for many of Adelbert Ames's famous perspective illusions.[2] The bizarre and disturbing 'Ames room' has sloping floor, ceiling, and walls, but appears to have a normal rectangular form when viewed from one particular position. An infinity of different 'Ames rooms' could produce exactly the same appearance from that viewpoint.

If, however, we are prepared to make some plausible assumptions about the general nature of the geometry of the scene depicted, then much of this potential perspective ambiguity disappears. We can safely assume in the case of Vermeer's paintings that the overall architecture of the rooms is rectangular; that walls are vertical and floors horizontal; that doors, windows, and tabletops are rectangles; and that the floor tiles are square. Furthermore, we know a great deal from our everyday experience about the solid forms of objects with non-orthogonal geometry, such as musical instruments or items of crockery. We use such knowledge and make these kinds of assumptions all the time, of course, when reading photographs and realistic pictures of all types.

In reconstructing the details of Vermeer's interiors there are a few instances where it is not possible to deduce the precise three-dimensional forms of the objects depicted from an intelligent interpretation of the painted surface: but these are minor difficulties.[3] There remains, however, one important general respect in which the spatial geometry *cannot* in principle be determined directly from measurements made on the surface of the painting, and that is the absolute *scale*. The pictures, speaking strictly geometrically, might equally well show dolls in dolls' houses as living people in normal-sized rooms. Naturally we read Vermeer's paintings as the latter (though I shall rely on this theoretical point later, when simulating the appearances of real full-size interiors by means of a reduced-scale model).

Even so, this would not generally fix the scale of the paintings precisely. We might assume that any male figure was about 1.8 m (6′) tall, any female figure 1.5 m (5′) say, the height of chair seats around 45 cm (18″), and so on. But all these dimensions are still variable within quite wide limits. With Vermeer, however, as perhaps with no other painter, because of his passion for visual accuracy, the situation is different. Chapter 2 described how all of the maps that appear in Vermeer's paintings are real printed maps, whose exact dimensions are known. The same is true of some of the pictures, by other artists, which Vermeer shows hanging on the walls of the rooms. If we assume that he depicted these at their precise actual sizes, then this means that we are able to fix the absolute scale of many of the paintings to within a few per cent. Other means of fixing the scale are provided by the musical instruments and by the Delftware tiles in the skirtings, which were manufactured to standard sizes.

All of Vermeer's interiors are 'frontal' or 'central' perspectives: that is, the plane of the picture is set precisely parallel with the far wall of the room. We know this because the images of the top and bottom of that wall are always horizontal in the picture, as are the ceiling joists and the top and bottom edges of any maps and paintings that hang on the wall. It follows that, where a second wall is visible at the left of the picture, this wall must be precisely at right angles to the picture plane. The result is that in each painting—so far as the architecture goes—there is a single *vanishing point*, at which the images of lines receding away from the viewer converge. Figure 29 shows these lines continued, to meet at the vanishing point, in *The Music Lesson*. (Any piece of furniture that is not aligned with the walls but set at an angle, like the chair just beyond the table, will have its own separate vanishing points.)

There are four steps in the process of three-dimensional reconstruction.[4]

1. The first step involves finding the central vanishing point of the picture, as described, and drawing a horizontal line through this point to mark the theoretical *horizon* (Figure 29).

2. The next step is to draw in the images of the diagonal lines made by the

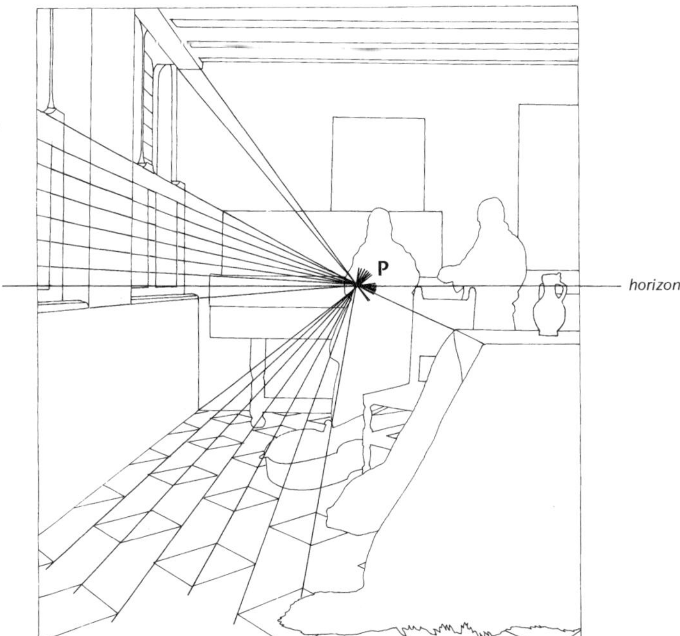

Figure 29
Receding lines converge to the central vanishing point P in *The Music Lesson*. The horizon line passes through the vanishing point.

patterns of floor tiles, and continue them to meet the horizon line. These diagonals also converge, in two points to right and left of the picture, which lie, like the vanishing point, on the horizon. These are the *distance points* (Figure 30: the figure shows only one of them). The distance points are spaced equally on either side of the vanishing point. The accuracy of Vermeer's perspectives is such that the two distances—which should in theory be exactly the same—never differ by more than a few per cent.[5] These values provide information with which to determine the position—in space, in front of the painting—of the picture's theoretical *viewpoint* (Figure 31).

This is the position at which Vermeer would have had to put his eye in order to see the scene in the precise perspective view represented in the picture. It is the position at which he would have had to place the lens of his camera obscura, if that was indeed how he worked. It is, very strictly, the point at which any viewer of the painting should place his or her eye in order to see it in correct perspective—although in practice we can look at most perspective pictures from positions well away from the theoretical viewpoint without noticing any

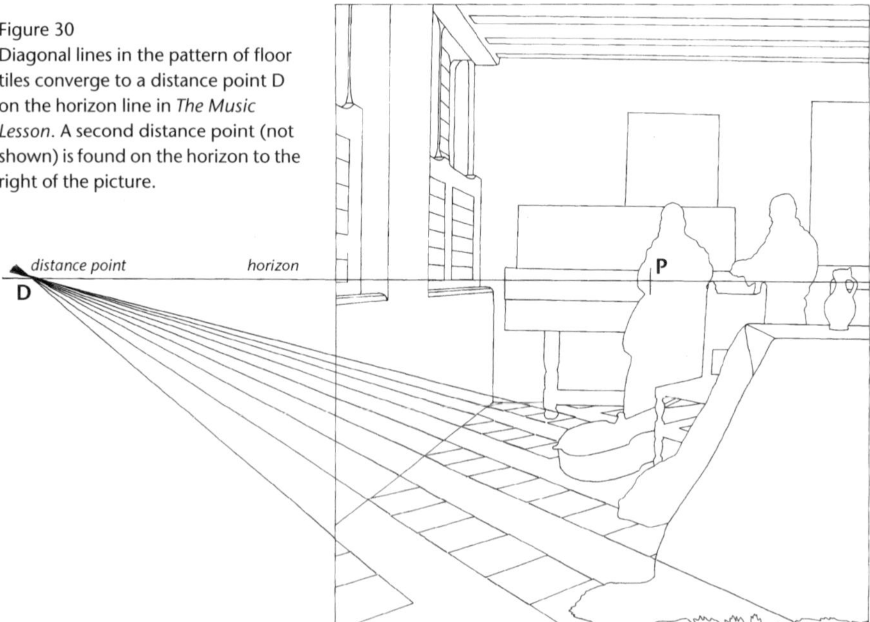

Figure 30
Diagonal lines in the pattern of floor tiles converge to a distance point D on the horizon line in *The Music Lesson*. A second distance point (not shown) is found on the horizon to the right of the picture.

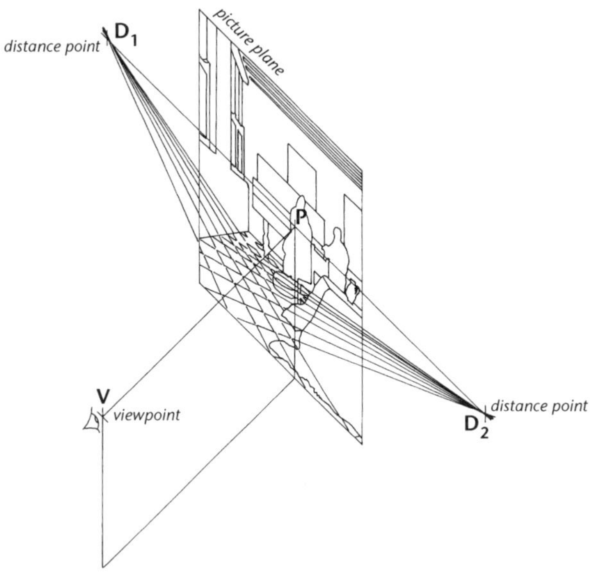

Figure 31
Position of the theoretical viewpoint V of *The Music Lesson*, on a line passing through the vanishing point P perpendicular to the picture plane. The distances VP, D_1P, and D_2P are all equal.

effects of distortion. The viewpoint in this type of perspective lies on a line perpendicular to the picture plane, passing through the central vanishing point. The distance of the viewpoint from the plane of the picture is equal to the distance between the vanishing point and either of the distance points. It will be clear from the description so far why a regular pattern of floor tiles is essential to this method of reconstruction.

3. Once the position of the viewpoint is established it is possible to move to the next step, that of working out a plan view of the room and furniture. The plan of the tile pattern is determined first of all. This is done by extending the pattern back from the line where the floor cuts the plane of the picture (see Figure 32). The sizes of the tiles can be measured directly at this floor line (labelled FF in the figure).

There are then two ways of finding the position in plan of some chosen point on any other object or feature appearing in the painting. Either the position can be found by means of its 'grid reference' in relation to the pattern of tiles—that is, by observing where some feature such as the foot of a chair-leg or the foot of a person falls on the floor, and counting the number of tiles from the edges of the room.

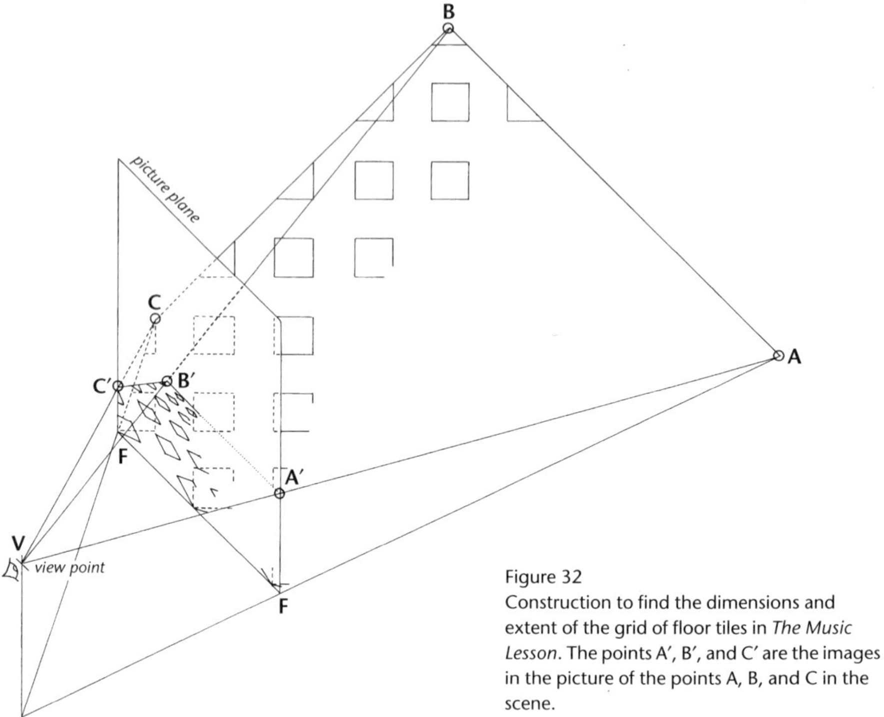

Figure 32
Construction to find the dimensions and extent of the grid of floor tiles in *The Music Lesson*. The points A′, B′, and C′ are the images in the picture of the points A, B, and C in the scene.

Alternatively a line can be carried from the viewpoint through the image of some feature in the picture plane to meet the position on the floor or wall surface where that feature is located. A line of this kind is called a *projector.* If the picture plane is imagined as a sheet of glass, then the projector can be thought of as a ray of light reflected from the object, passing through the picture plane, to meet the artist's eye at the viewpoint. Our process of reconstruction follows the path of this ray *backwards*, from eye, through picture, to object.

Thus the point A′ (referring to Figure 32) is the image in the painting of the point A at the edge of the floor in the room itself. We draw a straight line, the projector, joining the picture's viewpoint to this point A′ in the picture. We continue the projector beyond the picture to meet the floor plane, in order to determine the exact position of the point A in the scene. Figure 32 shows two more projectors BB′ and CC′, used to find the points B and C. With these the whole outline shape of the visible part of the floor can be determined.

4. The final step, that of drawing a side view of the room, is very similar to that for the plan. Again the positions of objects are determined by drawing projectors. For some features it is necessary to use both plan and side view together in order to find their locations.

Figure 33 shows a plan and a side view, constructed as described, for the space of *The Music Lesson*. The viewpoint is labelled V in each case. (Similar drawings for all eleven paintings for which reconstructions are possible can be found on the website associated with this book, at www.vermeerscamera.co.uk.) The plan shows only that part of the pattern of floor tiles which is visible. The positions of objects which are not wholly visible, but which can be located with confidence—for example the outline of the table beneath its covering drape, are shown in dotted lines.

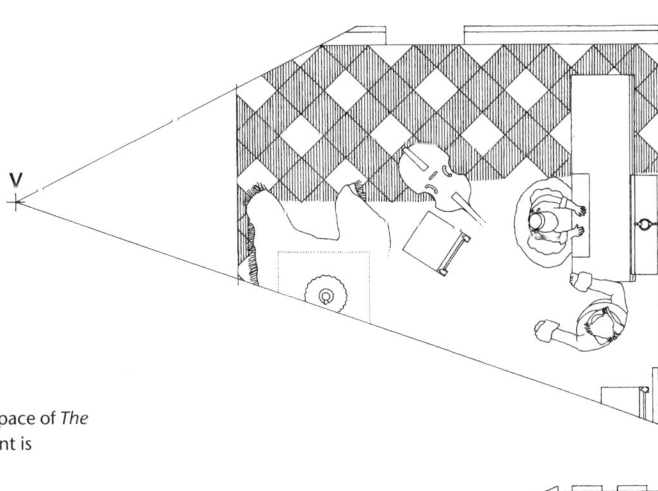

Figure 33
Plan and side view of the space of *The Music Lesson*. The viewpoint is labelled V in both cases.

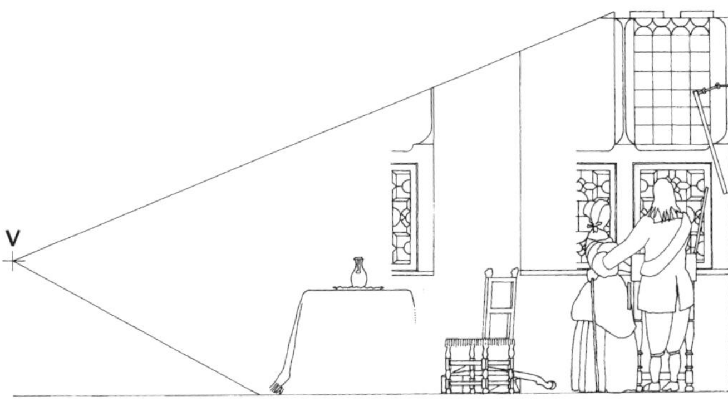

As mentioned in the last chapter, Swillens also made reconstructions of the rooms shown in several paintings and published plans and side views, very comparable to those illustrated here.[6] Examples of his drawings for *Officer and Laughing Girl* and *The Love Letter* are reproduced in Figure 34. From this analysis Swillens drew a number of general conclusions about Vermeer's working habits, which my own measurements confirm. Vermeer's viewpoint—that is to say the position of his eye—is always at or about the height of the window ledges, indicating that he sat to paint. It is rather rare that he sits directly opposite the centre of a composition. More often he is displaced to one side.

Swillens does not explain exactly how he made his reconstructions. It may be that a certain amount of guesswork was involved. For one thing he shows reconstructions of paintings, like *Officer and Laughing Girl*, in which no floor is visible, and of other paintings where the floor can be seen but is not tiled. With such pictures there is no reliable means of determining the depths of objects in the scene.

Again, although the floor is tiled, his reconstruction for *The Love Letter* has the viewpoint much nearer to the subject than it really is. As a result, Swillens has an angle between the two lines defining the visible area of the plan—the 'angle of view'—that is wider than it should be. The same is true of the equivalent angle in the side view. Figure 35 shows my own reconstruction for comparison.[7] One suspects that Swillens is trying to fit his preconception, that this painting shows a room of the same length and with the same pattern of windows as that seen in *The Music Lesson*. Thus he includes three windows in the side view of *The Love Letter*, where none can be seen in the painting. There are further completely conjectural elements included in some of his other reconstructions. By contrast, the rigorously measured geometrical procedures adopted here ought to produce much more reliable reconstructions.

One of my findings during the course of this geometrical work was that the pattern of floor tiles is painted with great precision in all cases. It never happens that incompatible parts of a pattern are found in discrete parts of the floor. That is to say, the pattern always joins up correctly behind obstructions. This contrasts, for example, with the perspective of floor patterns in Pieter de Hooch's work. Figure 36 shows *A Woman Drinking with Two Men*, where de Hooch has added an extra impossible row of floor tiles at the left, in order to create more apparent space for the woman and side-table. If this row of tiles is extended to meet the back wall, it is seen to fall outside the room—although this fact is

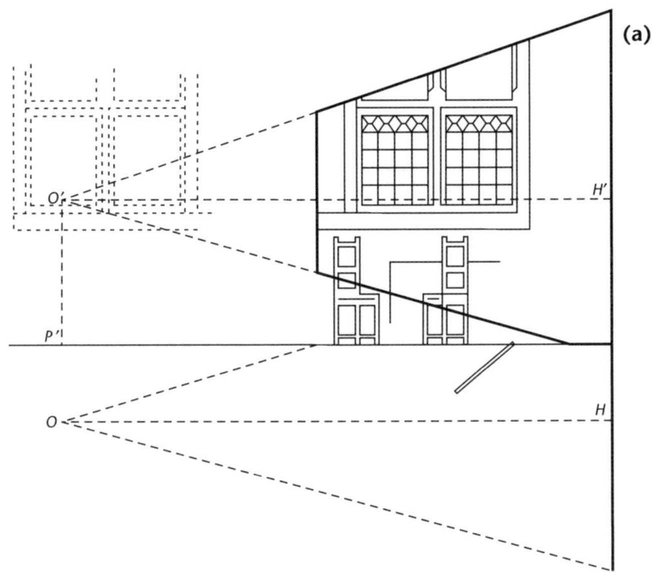

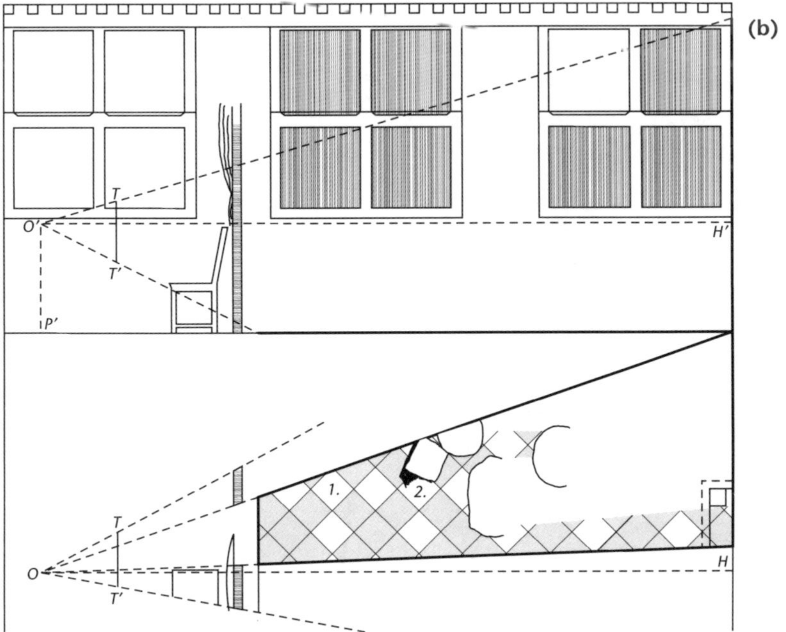

Figure 34 Plans and side views of the spaces shown in two paintings, as reconstructed by Swillens: (a) *Officer and Laughing Girl*; (b) *The Love Letter*.

Figure 35
Author's plan and side view of the space of *The Love Letter*.

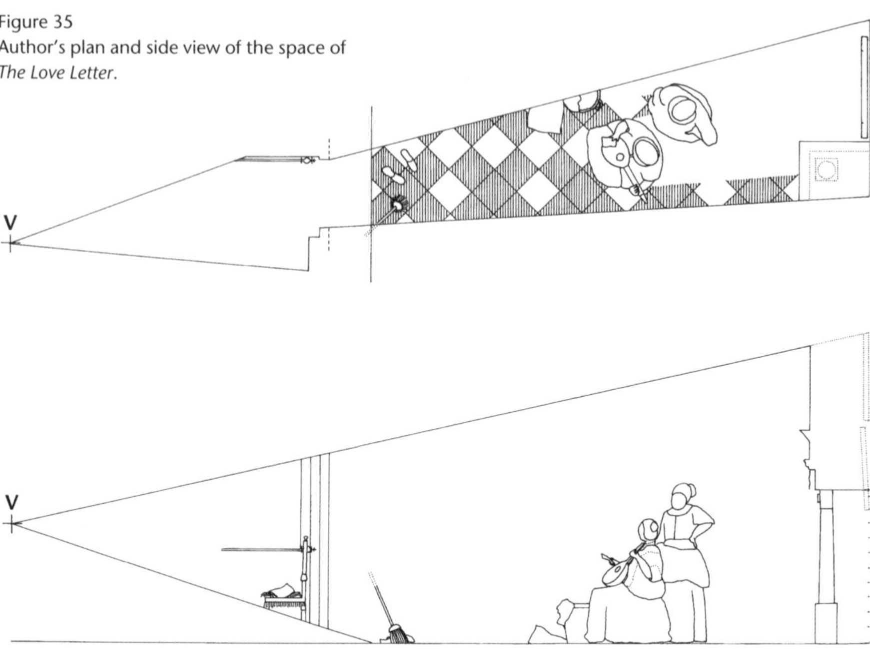

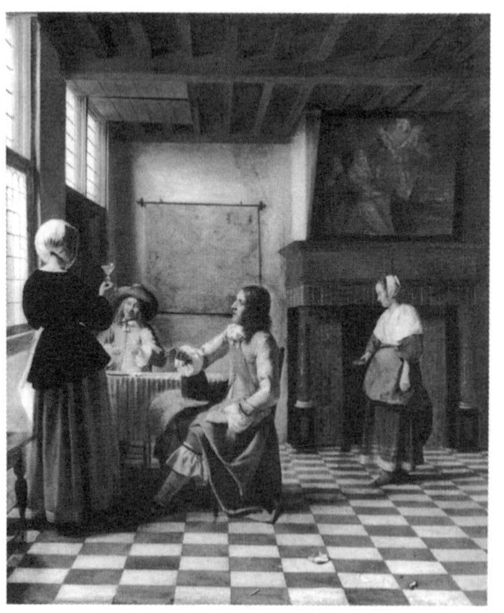

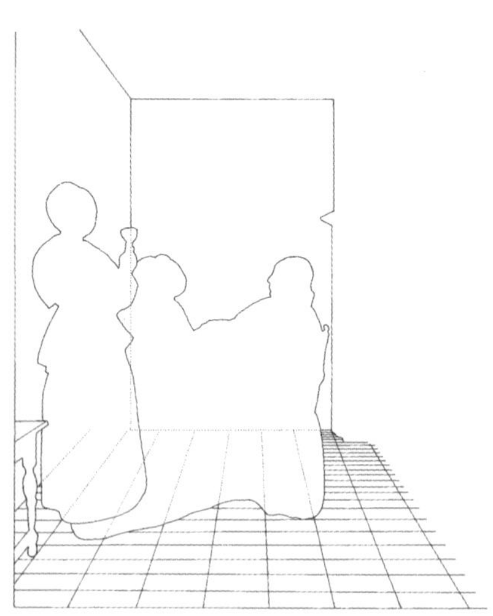

Figure 36 Pieter de Hooch, *A Woman Drinking with Two Men, and a Serving Woman*, c.1658, National Gallery, London. Oil on canvas, 73.7 × 64.6 cm. This shows an impossible row of tiles added at lower left.

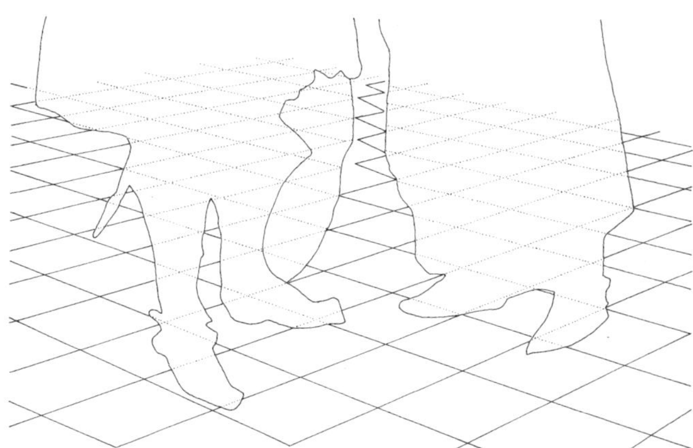

Figure 37
Outline of the tile pattern between the legs of the figures in Pieter de Hooch's *A Soldier Paying a Hostess* (compare Figure 14).

hidden by the placing of the larger table and the seated gentleman. Figure 37 shows the outline of part of the tile pattern in de Hooch's *A Soldier Paying a Hostess* (Figure 14) between the legs of the two figures. If the pattern of tiles is completed behind the figures (dotted lines) it becomes clear that de Hooch's detail does not conform to the correct perspective image of the grid. (Not that this is at all important to the way we read the pictures—such minutiae generally pass quite unnoticed.) These features of de Hooch's paintings argue for a compositional technique in which the perspective was sketched on the picture surface, and later altered or elaborated *ad hoc*. Such approximations are not found in Vermeer. Rather his geometrical consistency suggests either *extremely* careful perspective construction or else very close observation of real floors or their projected images.

One picture tells us more about the geometry of the room depicted than any other, and that is *The Music Lesson*. It tells us more because of the *mirror* hanging on the far wall, above the virginals. Figure 38 shows a detail. Besides the woman's face and the carpet-covered table, this mirror reflects part of the black and white tiled floor. These tiles end at the top left-hand corner of the mirror, and above them we can just make out a small area of the end wall—that is, the wall *behind* our viewpoint, behind Vermeer as he painted.

At the top right-hand corner of the mirror is the reflection of what is at first sight a puzzling group of pieces of furniture, not directly visible in the picture. Closer inspection, however, shows the nearest item to be an easel carrying a canvas. Beyond it are the legs of a stool. Between stool and easel

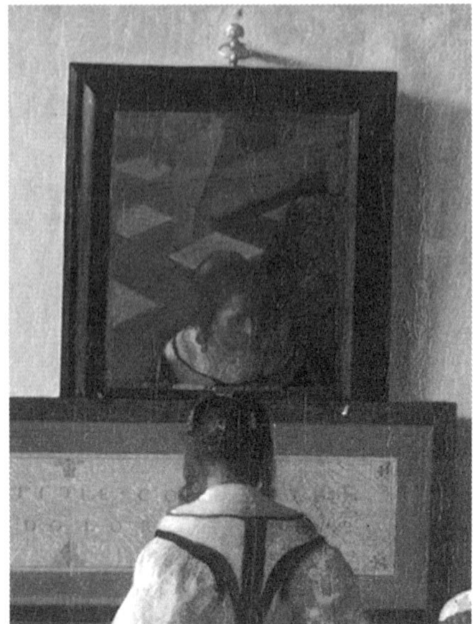

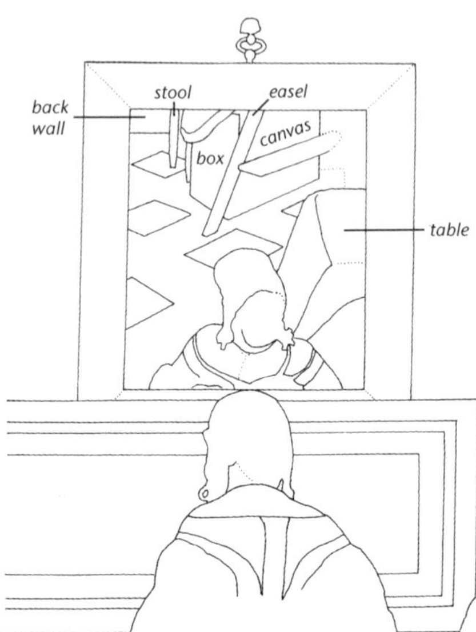

Figure 38 Detail of the mirror in *The Music Lesson*.

stands what is apparently some kind of box, although it is very indistinctly represented.

If we look again at the reconstruction of the space of the room in side view (Figure 39), we see that it is possible to determine reasonably accurately the angle at which the mirror must be hanging, in order that it reflects what it does. For example, two lines drawn from the viewpoint and from the further edge of the table to meet the image in the mirror of that same edge must make equal

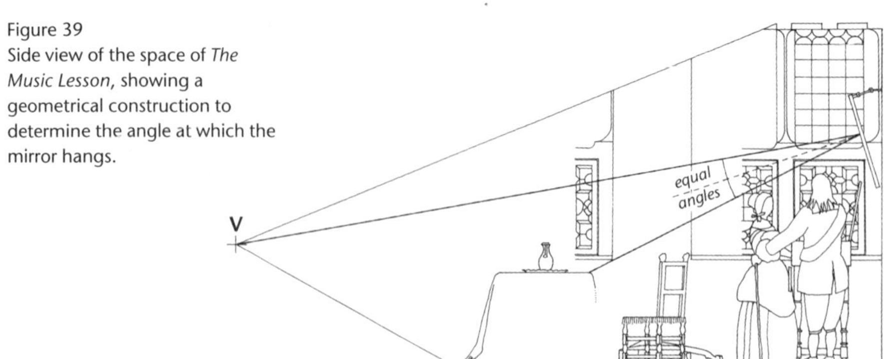

Figure 39
Side view of the space of *The Music Lesson*, showing a geometrical construction to determine the angle at which the mirror hangs.

angles of reflection with the mirror's surface. The same applies to the nearer edge of the table, and so on.

With this angle between mirror and wall established, it is possible to construct a plan of the part of the room that the mirror shows. Figure 40 gives a bird's-eye view of the room, including the area visible only in the mirror reflection. The stool, the easel, and the 'footprint' of the box are all included. The pattern of tiles matches up across the two areas of floor—the area that we see directly and the area seen in the mirror. It turns out that the room is the length of sixteen of the large marble tiles set diagonally. Both end walls cut through the centre of a row of white tiles (as does the side wall), so that the overall tile arrangement is neatly symmetrical in relation to the room's edges.

The estimated viewpoint of the picture lies above one corner of the stool seen in the mirror. It is indeed at the eye-position of someone seated on the stool. Can we perhaps catch a glimpse of Vermeer himself? Very disappointingly, there seems to be no sign of him at all, not even his boot or the edge of his jacket. As Benedict Nicolson says, 'the painter appears to have been swamped by a wave of diffidence and at the crucial moment to have fled the scene.'[8] The easel is set diagonally to the right of the stool, just as the artist has *his* easel in *Allegory of Painting*. It is positioned such that the canvas does not obscure, with its left-hand edge, any part of the painter's angle of view.

Notwithstanding Vermeer's shyness, it is clear that he has positioned the mirror extremely carefully in order to let us, the viewers, see his own vantage-point. It is an odd place to put a mirror in any case, over the virginals. Its angle and height on the wall are such that no vain young woman could look at her face in it from close up. (Compare for instance the placing of the mirror on the left-hand wall in *Woman with a Pearl Necklace*.) And yet, characteristically of this enigmatic and evasive man, Vermeer obscures as much as he reveals. Does he really mean us to imagine him painting in this position? Is it perhaps significant that no one sits at the viewpoint?

The fact that Vermeer lets us see the conventional tools of his trade at the very point from which he has painted the picture might seem, on innocent first thoughts, to dispose of any camera hypothesis. On the other hand, Gowing points to the curious box between easel and stool. What makes this detail particularly tantalizing, is that it hints, both by its position and the fact that Vermeer—one suspects deliberately—shows it clearly enough to intrigue, but not clearly enough to explain, that it might possibly have some important

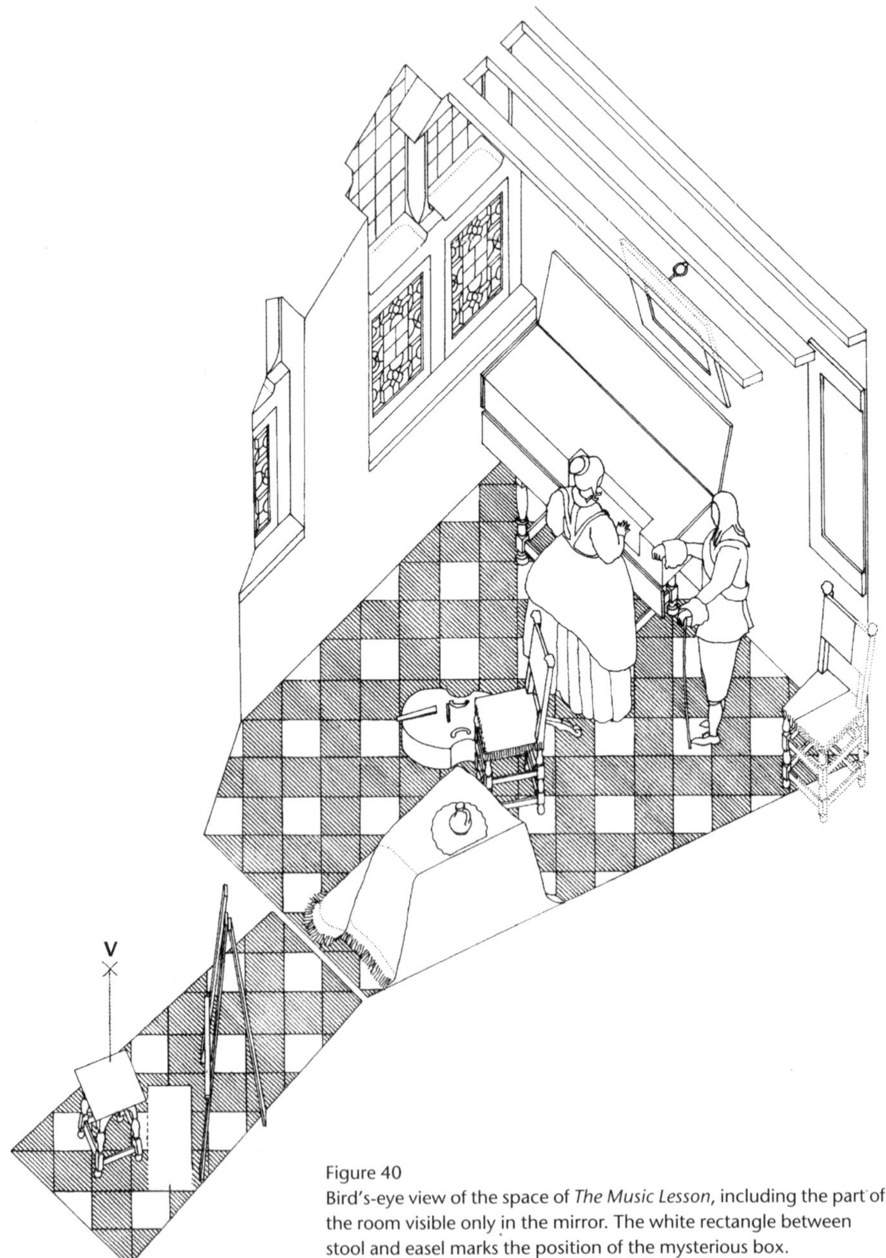

Figure 40
Bird's-eye view of the space of *The Music Lesson*, including the part of the room visible only in the mirror. The white rectangle between stool and easel marks the position of the mysterious box.

significance for his technique. As Gowing says, if *The Music Lesson*, as some people have suggested, indicates the way in which Vermeer worked, then this is 'hardly in the usual sense. To complete the story we should probably need to discover the nature of the box which stands behind the easel in the reflection, where we might expect to see the painter himself. Possibly his intimates understood the reference.'[9]

I have puzzled and puzzled over this question, but must confess to being ultimately defeated. It is perhaps worth recording my thoughts, all the same, in case others might be prompted to a solution. I have not even been able to decide the exact shape of the 'box'. At one time I thought it might have a sloping top—the shape perhaps of a small roll-top desk, if such a thing were not highly improbable, not to say wholly anachronistic. The alternative is that it has a flat top, and that the sloping line at upper left is the edge of a cloth or carpet placed over it. The upper part of the front face of the box is hidden behind the canvas on the easel. It is very difficult to see the canvas in most reproductions, but in the original it is possible to make out the stretcher with two of its diagonal wooden corner-struts. The back of the canvas itself seems to be painted in a dark reddish brown, in contrast to the more yellowy brown of the box and easel.

Is the box a camera? In 1685–6 Johann Zahn published a book, *Oculus Artificialis Teledioptricus, sive Telescopium* (*The Long-Distance Artificial Eye, or Telescope*), which illustrates a number of novel designs for portable or movable box cameras. Some of these seem to have distinct affinities in overall form with what we seem to see in Vermeer's mirror (Figure 41).[10] Zahn's cameras are tall, wide, and shallow; they have vertical lens tubes with 45 degree mirrors above; one even has a sloped viewing screen. This is immediately exciting, but on further reflection creates many more problems than it solves. If Vermeer's box *is* a camera, then it is neither of an appropriate design, nor in the correct position, to obtain an image of *The Music Lesson* as he paints it. First, it is not at the correct viewpoint of *The Music Lesson* in plan, which is a small distance further back, directly above Vermeer's stool. (Since we are unable to see the top of the box/camera, it is impossible to say whether its mirror might be at the correct *height* for the viewpoint of the picture. It could be.) Second, the box is not facing in the right direction, but is angled sharply to the right, as is the painter's easel, in order that he can see his subject beyond them. Third, the box is not sufficiently deep for a camera capable of producing an image at the size of the painting itself.

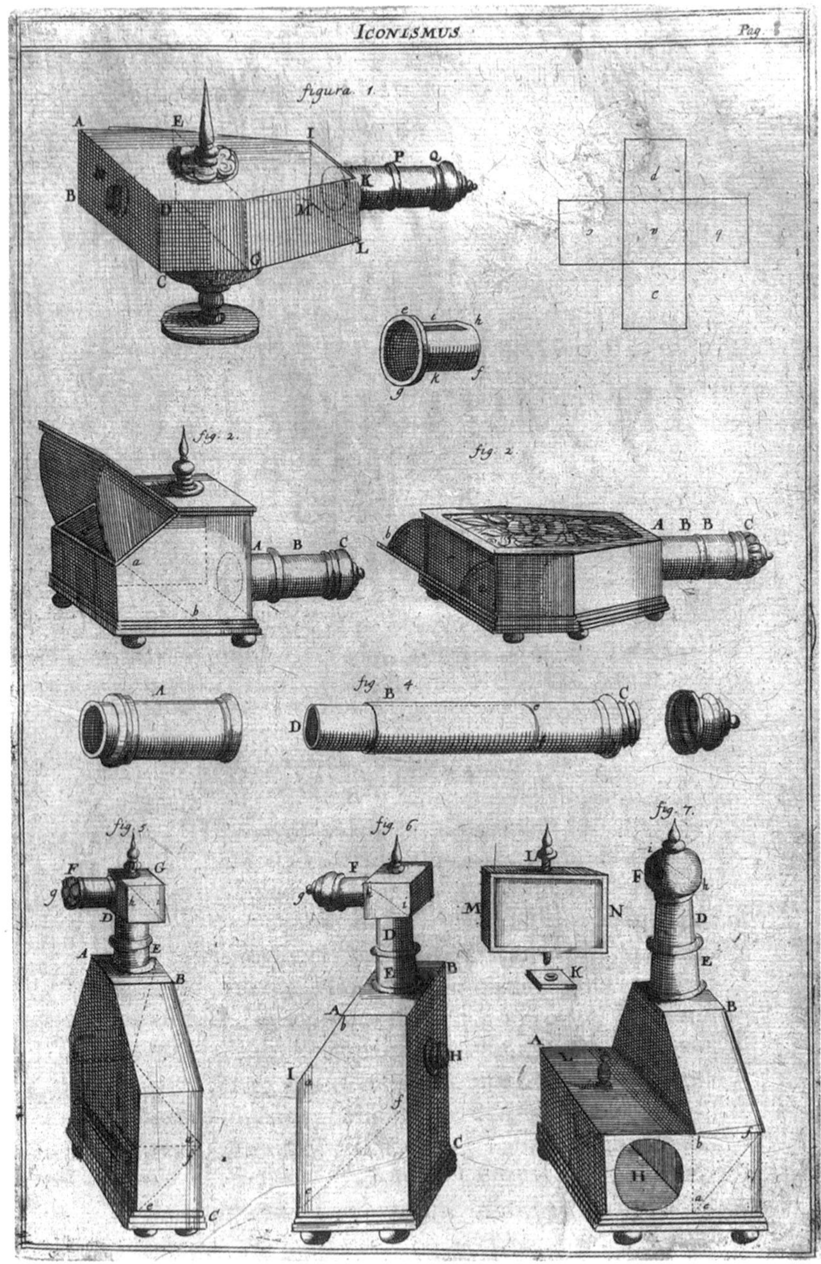

Figure 41 Designs of portable camera obscura from Johann Zahn, *Oculus Artificialis* (1685–6).

There is the further problem of why the camera—if such it is—should be placed *between* the painter and his canvas, completely obscuring the lower part of the painting from his view. In speculative mood, one might wonder whether this is possibly an extra, second camera, used in addition to some larger booth or tent. Perhaps in the later stages of painting, when Vermeer was working in full light in front of the actual scene, he might have had a smaller box camera beside his easel, giving a reduced-scale image to which he could refer from time to time. This would mean having the box camera in a quite different position from that in which Vermeer shows it in the mirror reflection. But all this is to put much greater interpretive weight than it can carry onto the end of a very long branch of unsupported conjecture.

It may be that, after all, the explanation is more prosaic. According to Swillens it was customary for 17th-century Dutch artists to store their precious pigments in chests 'with drawers and compartments'.[11] So perhaps what we see here is Vermeer's storage cabinet. Wheelock offers this suggestion.[12] There is no such box visible in Vermeer's own *Allegory of Painting*, but there are examples to be seen, set alongside easels, in several contemporary genre scenes of 'artists in their studios'.[13] Some take the form of little chests of drawers, others have hinged lids. As I say, I have abandoned the conundrum, reluctantly, at this point, with no more suggestions to offer.

This excursion into the mysteries of the mirror in *The Music Lesson*, though fruitless in some ways, has nevertheless provided one piece of information which will prove crucial shortly. The mirror allows us to see the back wall of the room in question, and so to measure the room's length. Let us return to the perspective analysis. Recall that the reconstructions of the spaces of the interiors are made, up to this stage, at varying arbitrary scales; and that no absolute dimensions are fixed. How can the true scales be determined? This is not simple in practice, and a great deal of trial and error is involved.

The sizes of any items of furniture and architectural features which are represented in the *same* painting are thereby necessarily all related. Suppose we assume some particular scale for the space shown in that one picture. It is convenient in practice to express this in terms of the assumed size of one floor tile. If we change our assumption—for example we increase the dimension of the tile—then all other objects or elements in the picture must increase in size proportionately. Similarly if the dimension of the tile is shrunk, all other objects shrink with it.

I made a first working assumption that any sufficiently distinctive features of the architecture of the room, and any recognizably identical pieces of furniture, are indeed the same size in all the pictures in which they occur. (That is to say, they have the same dimensions 'in reality', although their painted images may of course differ in size because of their varying distance in the scene.) It follows that the scale or scales decided for the respective paintings must be such that these dimensions remain constant throughout. One example is the type of chair with lions' head finials which features in three of the eleven paintings under consideration. The same chairs appear admittedly with different upholstery, sometimes blue cloth, sometimes black leather with painted yellow lozenges; otherwise the design remains identical. Another type of chair which recurs has a square back and is covered in tapestry or *petit point*, with a pattern of flowers and leaves in green and buff (see Figure 46). Two such chairs appear in *The Concert*. It seems to be the same design, viewed from the side, in both *Lady Seated at the Virginals* and *The Girl with a Wineglass*. The virginals in *Lady Standing . . .* and *Lady Seated . . .* seem to be the same instrument, on the basis of the very distinctive marbled case and black legs, but with two differently painted lids.

With other items the situation is not so clear-cut. Where tables appear they are often covered with carpets, so it is not easy to say whether it is the same table in different pictures. Nevertheless, there is a very characteristic design of heavy table leg with globular foot, rectangular section at 'thigh' and 'ankle', and fat vase-shaped moulding at the 'knee' (Figure 42) which appears in at

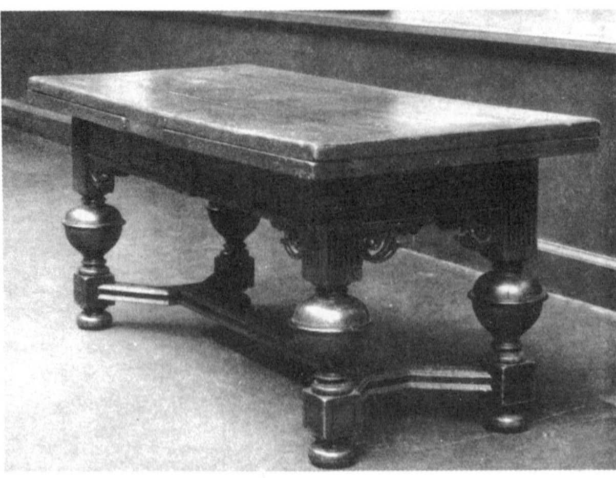

Figure 42
17th-century table in the Rijksmuseum, Amsterdam, with fat vase-shaped mouldings at the 'knees', similar to table legs visible in five of Vermeer's paintings.

least five paintings.[14] It would be reasonable to suppose that this leg supported the identical table in every case. Tables of this general design were relatively common in 17th-century Holland. The figure shows an example in the Rijksmuseum in Amsterdam, illustrated by Swillens.[15] There are similar tables in the Prinsenhof Museum in Delft.

Naturally we would want all the windows of a given design to have the same dimensions and spacing in every instance. We would want the heights of their sills above floor level to be consistent. It would also be interesting if the overall height of the room or rooms, where this can be measured, remained the same.

I set a further requirement, or desideratum, in fixing the scale or scales: objects whose dimensions are known exactly from independent sources of information should indeed come to their true sizes. In the pictures under consideration Vermeer depicts three real maps. The map of Europe in *Woman with a Lute* was published by Jodocus Hondius in 1613. Figure 43 reproduces a copy of the 1659 edition by Joan Blaeu of the same map, in all essential respects identical to the Hondius version.[16] It is missing the surrounding printed text that Vermeer depicts. In his painted version we do not see the top or right-hand edges of the map, but we can nevertheless determine the true size of the visible portion by seeing where the edges of the painting fall relative to the geographical detail.

The map of the Seventeen Provinces of the Netherlands in *Allegory of Painting* was published by Claes Jansz. Visscher, probably at some time during the 1590s. One copy of the central map section was found in 1962 (Figure 44), although without the borders showing topographical views of Dutch towns.[17] However, the central part in the painted image can be measured separately, of course, for comparison with the real map. The third map appears in deep shadow in the left foreground of *The Love Letter*.[18] Unfortunately for the present purposes, it is seen in receding perspective; it is very dark and indistinct; and the wall on which it hangs cannot be located precisely in plan. For all these reasons it is not useful for measurement purposes.

One identifiable real painting hangs on the far wall in both *The Concert* and *Lady Seated at the Virginals*. This is Dirck van Baburen's *The Procuress*, now in the Museum of Fine Arts in Boston (Figure 45).[19] In *Lady Seated . . .* the image of van Baburen's picture is cropped at top and right.

The viola da gamba which features in four paintings seems likely to be the same instrument throughout, and in any case the dimensions of different

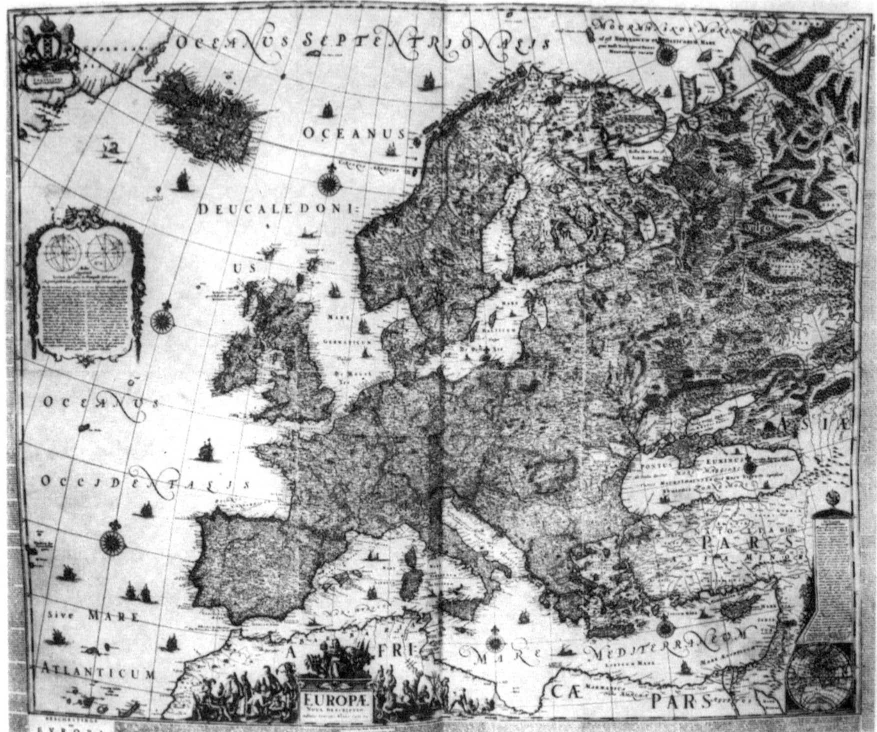

Figure 43 Joan Blaeu, map of Europe, 1659; first published by Jodocus Hondius, 1613. Vermeer reproduces this map in *Woman with a Lute*.

instruments do not vary greatly.[20] The virginals in *The Music Lesson* seem almost certain to have been made by the celebrated Antwerp firm of Ruckers. There is an almost identical instrument in the Musée Instrumental du Conservatoire in Brussels, signed by Andries Ruckers the elder.[21] Vermeer's virginals, including those in *Lady Standing . . .* and *Lady Seated . . .*, can be compared in size with actual contemporary instruments.[22]

As previously mentioned, in *Lady Writing a Letter, with her Maid*, and in both *Lady Standing . . .* and *Lady Seated . . .*, there are skirtings made from Delftware tiles, painted with figures representing trades and occupations. Most such tiles were made to a standard size, about 13 cm (5″) square.[23]

In finding scales there is finally the fact that certain dimensions of the room itself, in each case, must necessarily be interrelated. For example, take those

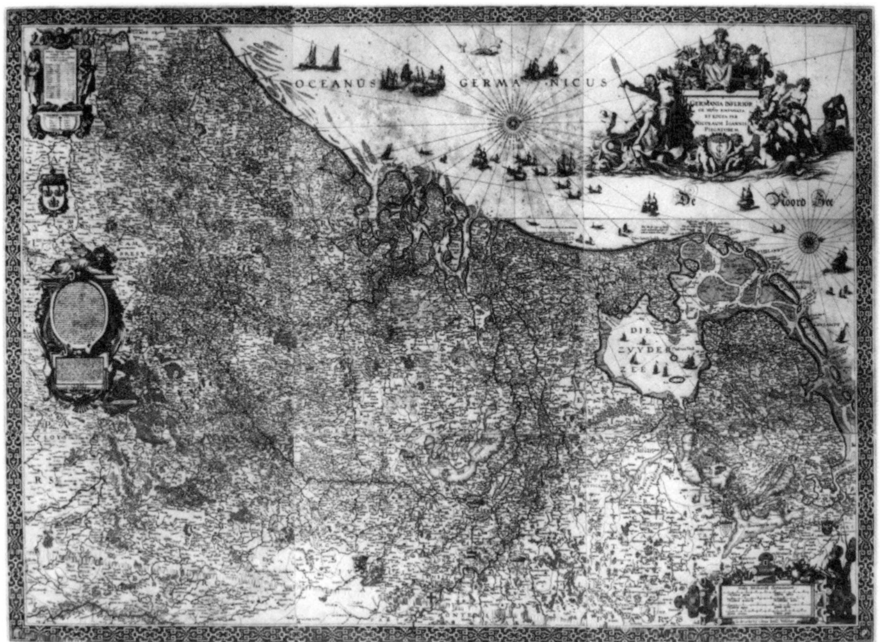

Figure 44 Claes Jansz. Visscher, map of Seventeen Provinces of the Netherlands, *c.* 1595 (central section only). Vermeer includes this map, ornamented with borders showing views of Dutch towns, in *Allegory of Painting*.

interiors where the windows are of the 'squares and circles' type, as in *The Music Lesson*. Each of the two visible window embrasures contains two casements: so the embrasures in their turn must be equal in width. Again we see the whole of two complete embrasures in *The Glass of Wine*—although here the casements in the further embrasure have their external shutters closed.

Furthermore, if these are rooms in a house or houses in Delft, then they are almost certainly *brick* houses. Brick construction imposes its own dimensional discipline, since the lengths of walls and the widths of openings in walls are in general made equal to some simple multiple of the length of one brick. This minimizes the number of bricks which need to be cut. If it turned out that estimated room dimensions were compatible with brick construction, then that would certainly lend a little more support to their credibility.

I decided to start, in all this juggling of sizes, by assuming as a provisional hypothesis that the black and white marble tiles are the same size in all the

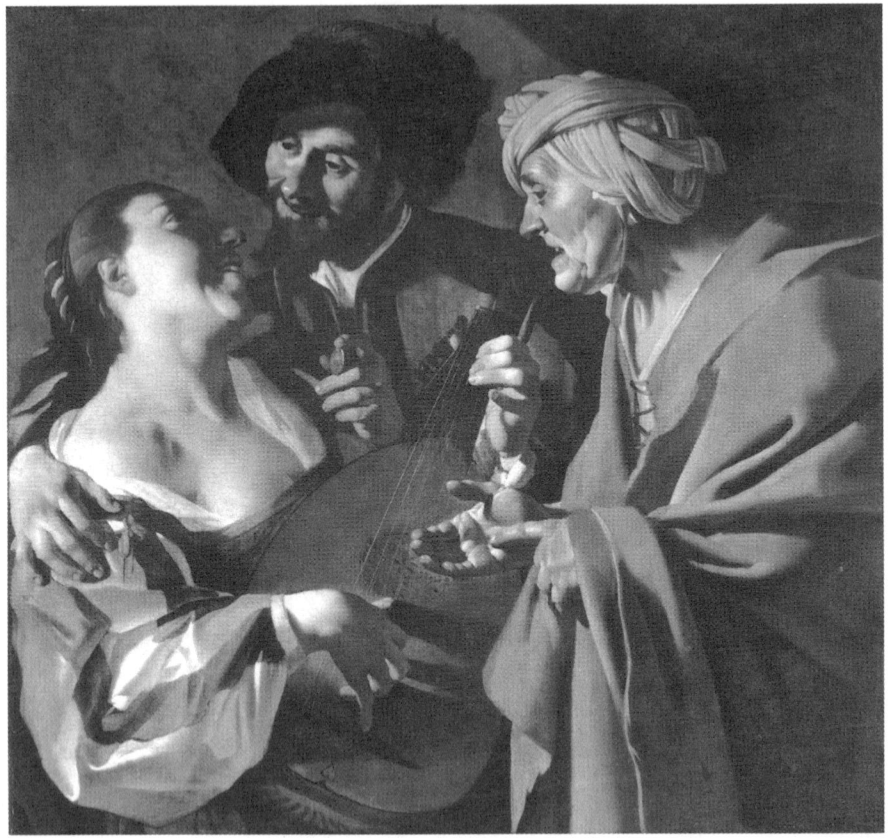

Figure 45 Dirck van Baburen, *The Procuress*, 1622, Museum of Fine Arts, Boston. Oil on canvas, 101 × 107.3 cm. Vermeer shows this painting in both *The Concert* and *Lady Seated at the Virginals*.

pictures in which they appear—despite their being arranged in varying patterns. I was therefore considering just nine paintings, and setting aside temporarily *The Glass of Wine* and *The Girl with a Wineglass*, where the floors are covered with the smaller ceramic tiles.

I was able to find by experiment that a standard scale based on a marble tile whose edge length is 29.3 cm (11.52") brings the two wall maps and van Baburen's *Procuress*, as seen in *The Concert*, very close to their true known sizes.[24] (Table B1 in Appendix B compares these actual sizes with their sizes as calculated from the reconstructions.) The visible part of the map of Europe is slightly larger (5%) in the reconstruction than in reality, and the central section of the map of the Seventeen Provinces slightly smaller (3 or 4%); the reconstructed *Procuress* differs somewhat in proportion from the real painting, being very close to the true width, but rather less than the true height.[25] All in all, however, within the limits of accuracy of the exercise, this particular standard scale represents a good compromise in relation to the known dimensions of maps and painting.

Swillens was able to identify the chairs with the flowered pattern of upholstery (Figure 46).[26] They are of the same design as a batch of 41 chairs delivered to the Town Hall in Delft in 1661 by the tapestry weaver Maximiliaan van der Gucht, for use by 'The Forty', the members of the Town Council (one chair to spare). A few of these are still kept in the Prinsenhof Museum.[27] Curiously they are not on public show, but hidden away in an attic. My colleague Marc van

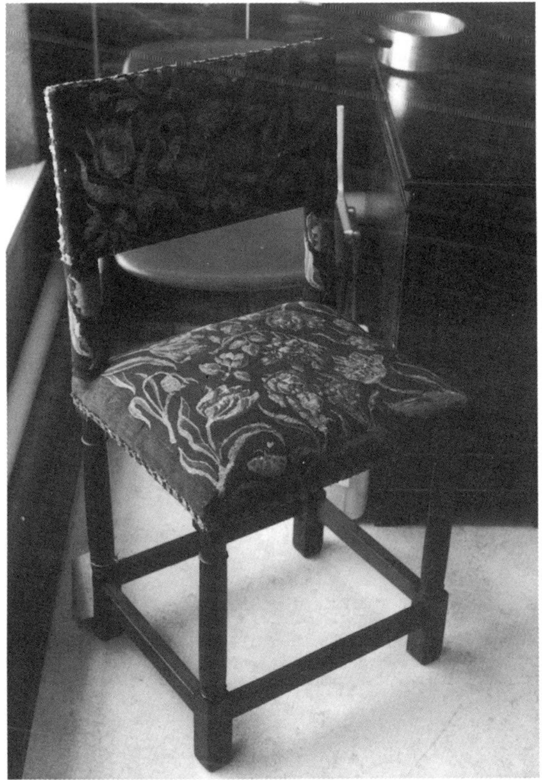

Figure 46
Chair upholstered in tapestry; one of a batch made by Maximiliaan van der Gucht for the Town Hall in Delft.

Leusen located and measured them for me. They turned out to have almost exactly the same dimensions as the reconstructed sizes of the chairs in *The Concert* and *The Girl with a Wineglass*. The reconstructions give 53 and 98 cm for the height of the seat and the total height. The real dimensions are 54 and 99 cm. There are also examples of the lions' head chairs in the Prinsenhof and in the Rijksmuseum in Amsterdam, whose dimensions are again within one or two centimetres of the average values calculated for these designs of chair from the reconstructions.[28] The calculations give 108, 42, and 42 cm for the height, width and depth of these chairs: the real dimensions are 108, 39, and 43 cm. (Table B2 in Appendix B gives estimated dimensions for all the furniture.)

What are the consequences of this for the sizes of other items in the reconstructions? The height of the standing man in *The Music Lesson* is now 1.8 metres (6′ 0″), while the heights of the standing women in this and four other pictures range from 1.55 to 1.62 metres (5′ 1″ to 5′ 4″), which seem very plausible. The average height of the tables is 76 cm (30″), which is entirely reasonable. This is the standard height for most tables and desks of any period. The Delftware skirting tiles come out at almost exactly 13 cm (5″) square, their standard manufactured size, in all three paintings in which they appear.

All this seemed very promising. I went on to try to reconcile the two pictures where the floor is made of ceramic tiles, with those where the floors are of marble. Ideally I wanted to find a scale, by fixing a size for the smaller tile, such that recurrent features like the casements and the sill height, and pieces of furniture like the lions' head chairs and the table with the bulbous legs, had the same dimensions as in the other nine paintings. It turned out that a ceramic tile which has exactly *half* the edge length of the marble tile—14.6 cm (5.76″) square—has just this desired effect. The evidence is presented in Tables B2 and B3 in Appendix B. There could be a special reason for this particular relationship between the two tile sizes, which we will come to in due course.

The data in Appendix B demonstrate convincingly that where recognizable pieces of furniture or other items appear in several pictures, their dimensions remain constant, within reasonable limits of accuracy. A more detailed discussion of these individual items, and of some minor anomalies that arise in the sizes calculated for them in different paintings, is provided in notes to the tables. The most serious of the discrepancies is that the architecture and the items of furniture are generally somewhat larger in *The Music Lesson* than in other pictures. I have no explanation. (In order to reduce the scale of the room and everything in it in just this one case, it would be necessary of course to break the assumption of a standard-size marble tile. I have been reluctant to do this, especially since *The Music Lesson* shares so many recognizable elements—casements, ceiling, chairs—with other paintings.) Otherwise 10 of the 11 reconstructed rooms, including those with the ceramic-tiled floors, begin to look more and more like the very same space. The sole exception is *The Love Letter*. Here the only feature shared with other pictures is the marble floor, and there are no visible windows, ceiling, or recognizable pieces of furniture. Consequently the analysis in this instance must remain inconclusive.

I raised the question earlier as to whether the dimensions of the architecture of the room might be governed by a brick module. To provide an answer we need to anticipate some of the findings of the next chapter, in which—out of a synthesis of all the measurements obtained here—a regular pattern for the windows is determined, along the whole of the wall in question. The pattern is shown in side view in Figure 47. For the moment I ask the reader to take on trust the hypothesis of a *third* window at the back of the room, identical to the other two, but never seen directly in any painting.

The bricks used in Holland in the 17th century were much smaller than modern bricks—typically only 16 or 17 cm in length. Unlike the Delft tiles, they were not standardized, being made in varying sizes by different brickworks, or

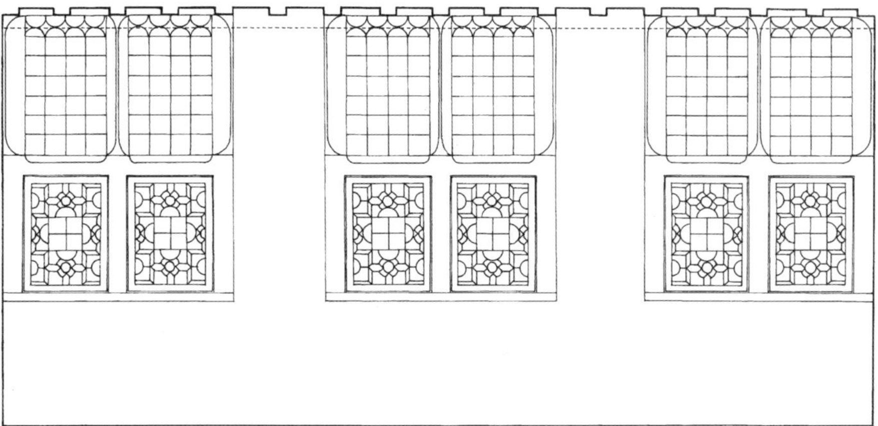

Figure 47 The left-hand wall of the room(s) with the 'squares and circles' pattern of windows. The third window at the left is, at this point in the argument, completely hypothetical.

fired to order on individual building sites.[29] The heavy wooden frames that form the basic structure of the windows in the reconstruction of Figure 47 are made of timbers with a 16 centimetre square section. This suggests that the frames were sized to fit a typical Delft brick module. Assume that the length of a brick (plus mortar) is precisely 16.2 cm. Then if each embrasure is exactly eleven bricks wide and each pier is exactly four bricks wide, these give dimensions very close to those obtained in the reconstructions. Such a brick pattern does *not* fit with the spacing of the joists in the ceiling, which have a different module. But then the joists are mortised at their ends into a timber wall-plate, and are not carried by the brickwork directly, so joists and bricks do not have to fit together.

I would not want to place too great a weight of significance on these brickwork calculations. There is plenty of margin for error here, and it is not too difficult to juggle the sizes to fit some preconceived arrangement. But the fact remains that the sizes determined for the architecture of the room *are* compatible with brick construction, on a plausible modular dimension.

Finally, after this long excursion into perspective geometry and measurement, we can come back to the questions raised at the beginning of the previous chapter, about the location of the room. Was it in Mechelen or in Maria Thins's house (or neither)? Estimates on the basis of the reconstruction of *The Music Lesson* show that the overall length of the room was 6.6 metres (21′ 10″). There may be some debate over the accuracy of the 1649 and 1675–8 pictorial maps of Delft,[30] but these are not the only maps to show the two buildings. The staff of the Cartography Department at the Technological University Delft were kind enough to locate for me a map made for property taxation purposes in 1830. The relevant area is reproduced in Figure 48, with the plans of Mechelen and Maria Thins's house highlighted. Because of the fiscal function and late date of the map, it can be expected to be dimensionally accurate.

The overall width of Mechelen according to this map is 7 m. Allowing for the thicknesses of the side-walls, this is nicely compatible with the length of the room of 6.6 m. It could have run across the whole width of the plan. Note from the map that this is not a standard width for Delft houses. Mechelen had a wider frontage than most of its neighbours, which tend to be between 4 and 6 m wide. Among 17th-century houses in Delft it is those with wider frontages like Mechelen which tend to have three windows per floor.

Unhappily for this line of argument, the Schrenk engraving (Figure 21) is

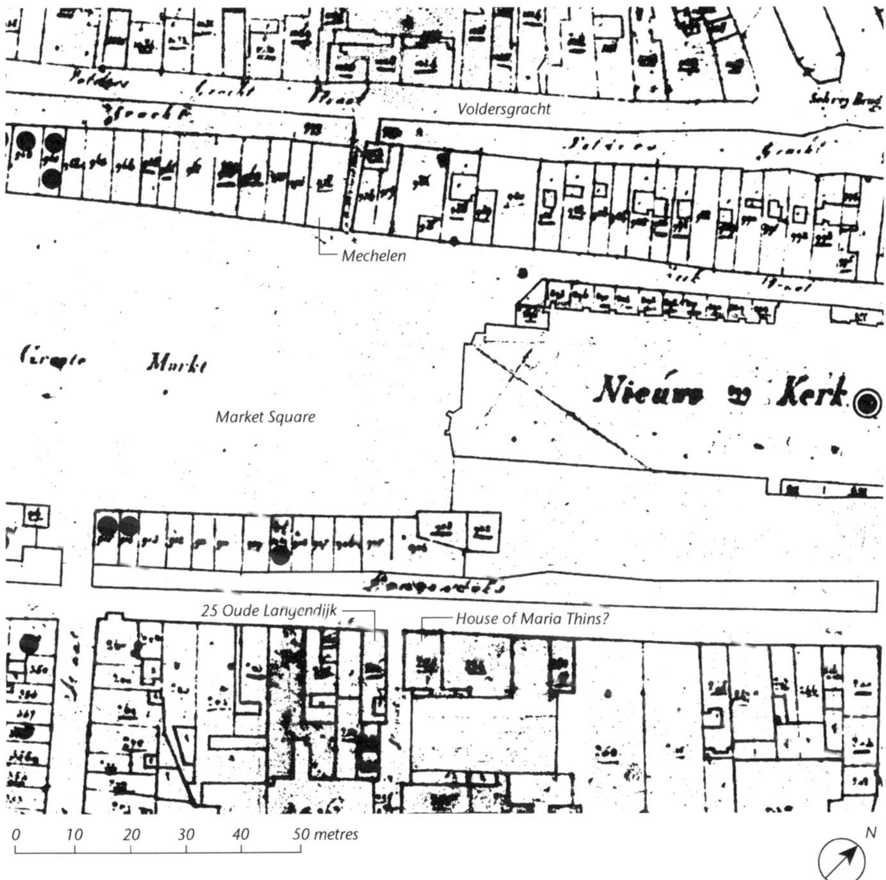

Figure 48 Detail of a map of Delft made for property taxation purposes in 1830, showing the Market Square and surroundings. Three buildings are highlighted: Mechelen; 25 Oude Langendijk, identified wrongly by Swillens as the house of Maria Thins; and the house of Maria Thins as identified by Montias on the opposite side of the Molenpoort.

unequivocal in showing *two* windows per floor on the main façade fronting onto the Market Square. This obviously weakens the case, although one might argue that a general topographical view of this kind—of which Figure 21 is a tiny detail—might not have been accurate to such a level of detail as the exact window pattern of every house. The view as a whole departs from the actual scene—which is not greatly changed today—in a number of respects. Perhaps more relevant is the point that we might expect Vermeer to have chosen a studio with north light—which is how his interiors appear to be lit—and that this is the south-facing front. Unfortunately there seems to be no surviving view of the north-facing rear facade of Mechelen, onto the Voldersgracht.

What about Maria Thins's house, as identified by Montias? I am unable to say with certainty whether the corner plot, indicated on the 1830 map, is or is not the actual plan of this house. (It seems not to extend as far back as the house in the 17th-century maps.) But suppose that the plot lines, at least at the front of the house, are the same as those of Maria Thins's original dwelling: then the width of the north-facing façade is again 7 m, like that of Mechelen.[31]

One other fragmentary clue is perhaps provided by the occurrence of windows of different types, in relation to the *dates* of the respective paintings in which they feature (compare Appendix A). The windows with the 'lozenge' pattern of leading are all found in pictures thought to have been painted in the late 1650s. Those with the 'squares and circles' pattern are only found in paintings dating from 1658 to 1670. This *might* suggest that the room with the 'squares and circles' windows was the front room of Maria Thins's house, given that Vermeer and his wife seem to have moved there some time before 1660.

It is only possible then, on the evidence offered here, to draw some tentative and dubious conclusions. Both Mechelen and Maria Thins's house are gone, and there are no surviving pictures of their north-facing fronts in which the very windows of Vermeer's putative studio might be recognized. On the other hand, the plans of both houses seem to be just about the appropriate width to accommodate a room of the correct dimension. And that is about all that can be said.

6 The riddle of the Sphinx of Delft

With the one difficulty of the somewhat larger sizes in *The Music Lesson*, the accumulated evidence from the ten paintings analysed in the last chapter is overwhelming: on the basis of what we can see of windows, walls, floor, and ceiling, the rooms shown in the various interiors have very similar dimensions throughout. Once the sizes of the marble tiles are standardized, the dimensions of other recurring features—where these can be measured with confidence—are brought close together. When the dimensions of the ceramic tiles are made half those of the marble tiles, the architectural dimensions in the two relevant paintings are reconciled with those measured from the other pictures with the marble floors. In this process the sizes of pieces of furniture and other 'props' are all reasonably standardized between pictures, and brought very near to their actual sizes where these are known. It is extremely hard to resist the conclusion that all these pictures do indeed show one and the same room. The changing tile patterns still present something of an obstacle to this view. It seems on the face of it as though this is a single room, but one in which the floor surface undergoes frequent changes. I will return to this question shortly.

We have now reached the crux of my entire argument. Suppose that we draw a plan of the room, and locate the viewpoints of the different paintings. Figure 49 shows such a plan, with viewpoints marked for six pictures: *The Girl with a Wineglass, The Glass of Wine, Lady Writing a Letter, with her Maid, Lady Standing at the Virginals, The Music Lesson*, and *The Concert*. The back wall, behind the viewpoints, is drawn in the position calculated from the mirror reflection in *The Music Lesson*.

The plan shows the angle of view of each picture, in the horizontal plane. If

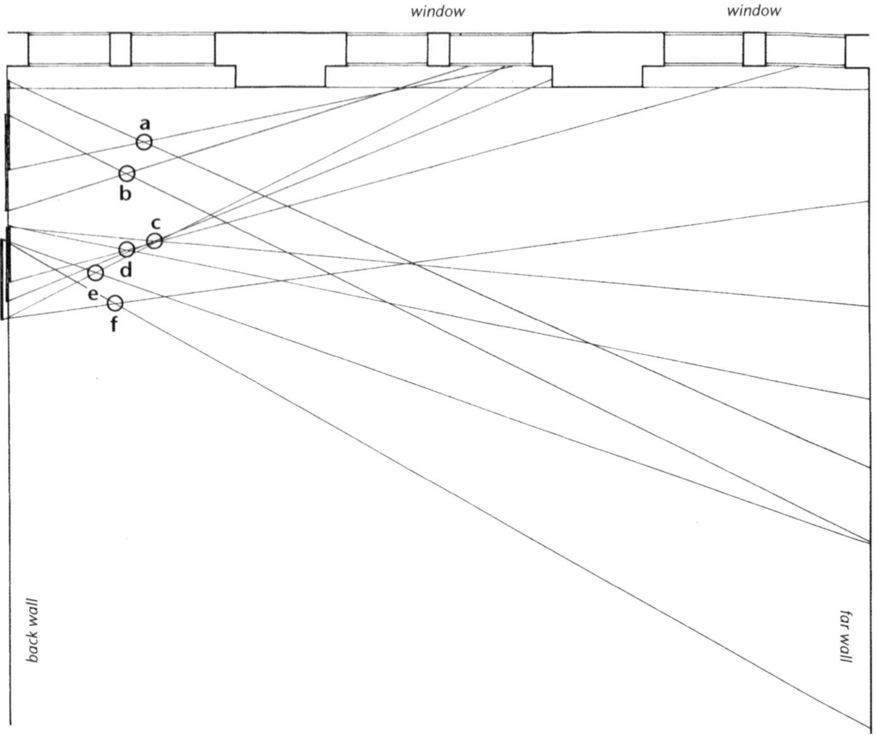

Figure 49 Plan of the room with viewpoints marked for six paintings: (a) *The Girl with a Wineglass*, (b) *The Glass of Wine*, (c) *Lady Writing a Letter, with her Maid*, (d) *Lady Standing at the Virginals*, (e) *The Music Lesson*, (f) *The Concert*. The diagonal lines mark the extent of what is visible in each picture. The third window is (still) hypothetical.

the two lines defining this angle of view are continued back, through the viewpoint, they meet the back wall at two points, as indicated. The same is true for the angle of view in the vertical plane, as seen from the side (Figure 50). By this construction we define, in effect, a rectangular area on the back wall. *In all six cases shown, this rectangle is almost exactly the same size as the relevant painting.* Figure 51 compares the actual dimensions of each painting with the dimensions of the corresponding rectangle on the wall.[1]

How are we to interpret this extraordinary geometrical coincidence? (It can hardly be due to chance.) It can be very simply explained by the assumption that Vermeer was using a camera. The lens would have been at the theoretical viewpoint in each case. The horizontal and vertical angles of view define the

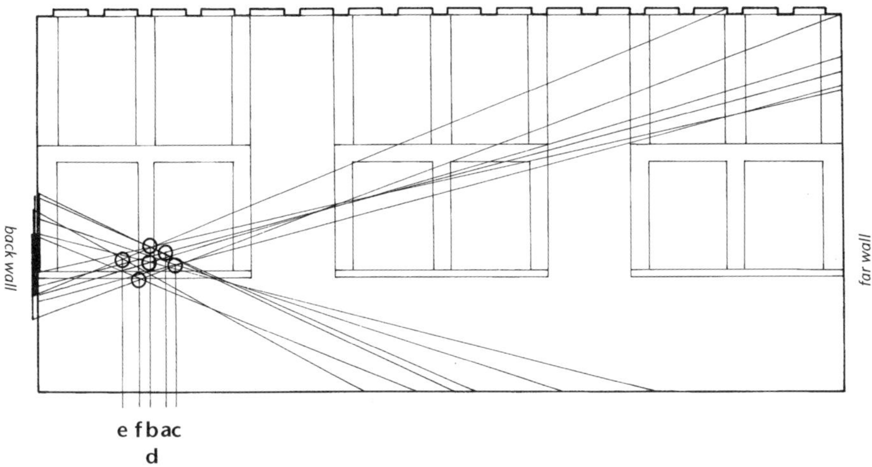

Figure 50 Side view of the room, with viewpoints and angles of view marked for six paintings, as in Figure 49: (a) *The Girl with a Wineglass*, (b) *The Glass of Wine*, (c) *Lady Writing a Letter, with her Maid*, (d) *Lady Standing at the Virginals*, (e) *The Music Lesson*, (f) *The Concert*.

limits of what is seen in the painting. The rectangle on the back wall marks the boundaries of the image of this scene, projected onto the wall by the lens. We can think of the diagonal lines in the diagrams as rays of light, passing from scene through lens to wall, marking the limits of each image. The geometry is that of the simplest room-type of camera, as illustrated diagrammatically in Figure 2. Each painting is the same size as its projected image because Vermeer has traced it.

In practice this would mean that Vermeer had some means of closing off and blacking out the back part of the room. He would then have worked inside this cubicle—a camera obscura in the original literal sense. Figure 52 gives an indication of what it might possibly have looked like. Notice in Figures 49 and 50 how the viewpoints of the six pictures are not all in exactly the same place. Vermeer shifts backward and forward some 50 cm or so, and moves laterally over a larger distance. This implies that it was possible to move the lens. Perhaps the front of the camera was made of a light screen, or curtains, rather than a solid wall. The lens might have been fixed in a tube poking out through these curtains. The depth of the cubicle is approximately 90 cm (varying slightly between pictures)—not large, but quite sufficient to sit and work inside in comfort.

Figure 51 Actual sizes of six paintings compared with the sizes of their 'projected images' as shown in Figures 49 and 50.

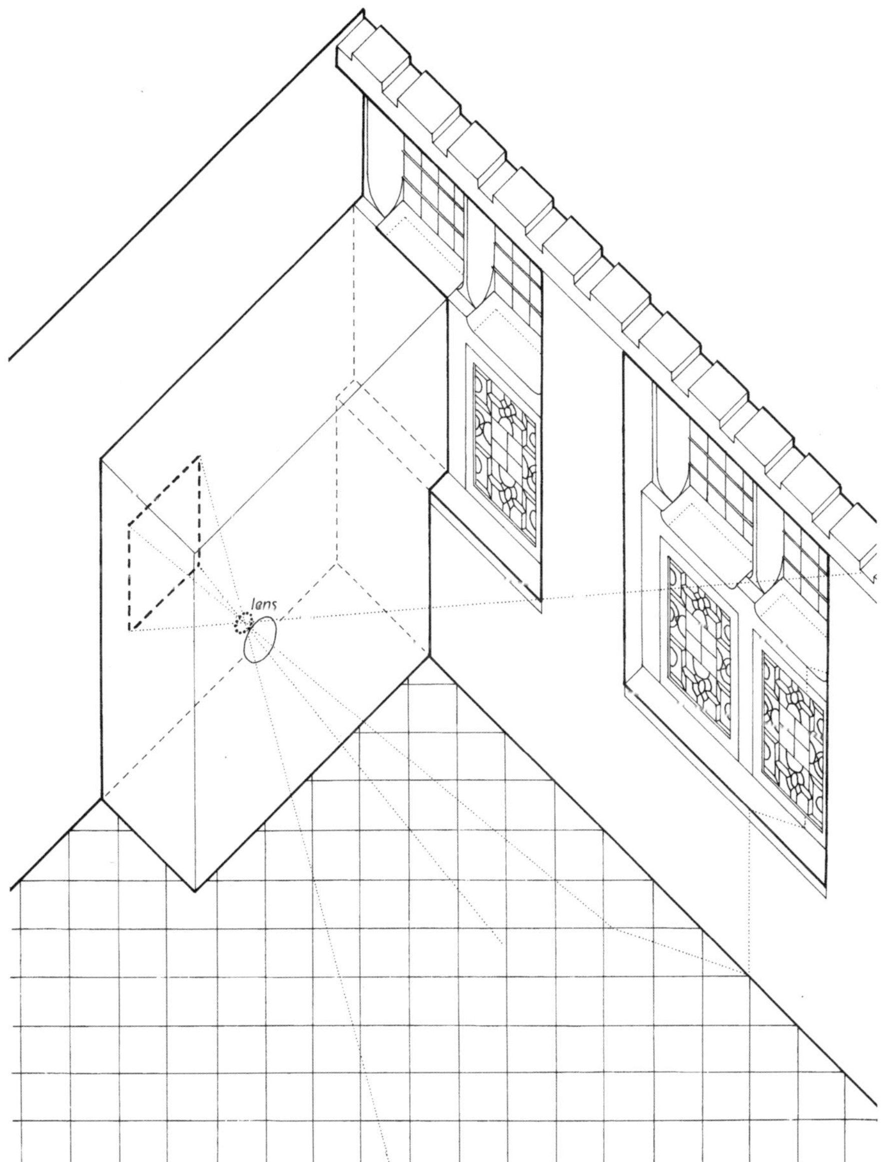

Figure 52 Possible arrangement for Vermeer's camera obscura.

The focal length of the lens can be estimated from the distances between lens and projected image, and between lens and subject, using the standard formula.[2] We can reasonably assume for this purpose that the subject—i.e. that part of the scene which would have been in sharpest focus—is where Vermeer's male and female figures sit or stand: usually somewhere opposite the nearer edge of the far window. On this basis the focal length of the lens would be something around 75 cm (30″). I will come back to the question of what diameter the lens might have had.

The complete extent of the image projected onto the wall in such a camera is circular in shape. We must imagine that Vermeer chose a rectangular area within this circle for each picture. The fact that the angles of view in Figure 49 are not symmetrical—more is seen to the right than to the left, or vice versa—implies that Vermeer did not in general select a rectangle at the very centre of this circle. (Had he done so, the vanishing points would all be at the dead centres of the paintings.)

It might be thought, from the way I have argued so far, that this geometrical finding depends absolutely on knowing the position of the back wall of the room from its reflection in the mirror in *The Music Lesson*. (Certainly it was via that piece of information that I made the discovery.) However, this is not the case. It would be possible to carry back the angles of view of the six paintings and find the plane in each case at which the 'projected rectangle' was equal in size to the painting itself. All these planes would then be found to coincide in one plane. The fact that this is *also* the plane of the back wall as calculated from *The Music Lesson* is of course a strong clue to a possible explanation.[3]

What of the five pictures, from the 11 reconstructed in the last chapter, whose viewpoints are *not* shown in Figures 49 and 50? The viewpoints of *The Love Letter* and *Lady Seated at the Virginals* lie outside the room altogether, well behind the back wall.[4] The viewpoints of *Allegory of Painting* and *Allegory of the Faith* fall just inside the room, but their 'projected images' on the back wall are very much smaller than the paintings themselves. *Woman with a Lute* presents special problems, although it is possible that this picture *does* have a 'projected image' on the back wall that is the true size of the painting. I will discuss each of these five pictures and their projected images in more detail later.

The use of the camera obscura is the most obvious immediate interpretation of this very curious characteristic of the combined perspective geometry of the six paintings; however, we should not overlook the possibility that other compositional techniques, optical or otherwise, might have been the cause. Some alternatives are explored in Chapter 8. To stay for the time being with the camera explanation however: can we say more about this darkened booth, inside which it seems Vermeer might have been working? And can we guess what such an arrangement might have involved for his painting technique?

Curiously enough, there is one other painting, besides *The Music Lesson*, in which we catch a reflected—if distorted and shadowy—view of the back of the room, behind our vantage-point. This is *Allegory of the Faith*, one of whose most striking features is the large mirrored sphere hanging above the seated woman (Figure 53a). Like several other items—the globe, the apple, the crushed snake—this is introduced to serve the painting's allegorical programme. It represents, in Wheelock's words, 'the ability of the human mind to contain the vastness of belief in God.'[5] Together with other details, this feature can be related directly to Cesare Ripa's *Iconologia* (*Symbols*) which provided an extensive repertory of images for painters to use in representing abstract ideas. A Dutch edition of the book was published in 1644, and Vermeer evidently followed it when planning his allegory.[6]

For our purposes, however, the iconographical significance of the ball is secondary: what is more important is the reflection on its surface, which Vermeer has painted as always in faithful detail (albeit in *very* soft focus). It is difficult, certainly, to identify some features of this image just by studying the painting itself, but in the next chapter we will see how this reflection can be recreated photographically, using a miniature mirrored sphere in a model of the room itself (Figure 53b). From this simulation it is possible to work out that the light-coloured patches at the bottom of the sphere are the reflections of the woman's face, white skirt, and sleeves, and of the Bible on the table. Above this is a reflected portion of the black and white tiled floor. And at centre left we see reflections of the windows of the room, which are not visible in the painting itself.

The reflection of one window is clearest. We see three lights—the two upper fixed lights and one of the casements—making an inverted L-shape. The second casement is presumably shuttered. To the left of this there is the image of another window, which is almost entirely curtained. However, there are two clear patches of sky where the curtains do not reach to the tops of the fixed lights. Also, the curtains themselves are not completely opaque, so that the light shows through as two slightly blueish areas against the blackness.

If we now look to the *right* of the inverted L-shape window, we see another

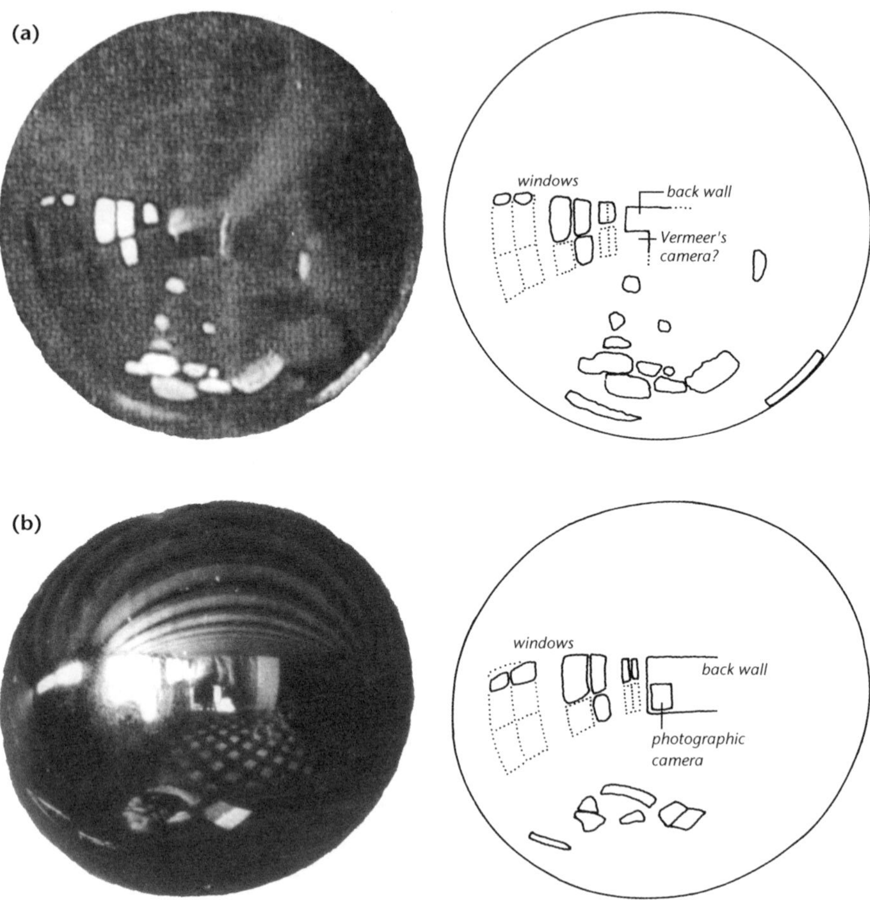

Figure 53 (a) Detail of mirrored sphere in *Allegory of the Faith*; (b) a photographic reconstruction of the same detail using a miniature mirrored sphere in a model of the room.

rectangular patch of brightness. This must be the upper light or lights of a *third* window, at the back of the room, adjacent to the back wall. Here is the third window whose existence was anticipated in the last chapter, in the reconstruction of Figure 47. We cannot see it in any other painting, since it is behind the viewpoint in every case.[7] There is no light coming from the casements in the lower part of this hypothetical window—supposing the design to be similar to the other windows.

Just to the right again in the reflection in the mirrored sphere is another roughly rectangular light-coloured patch, not as bright as the lights of the windows themselves. This must be the upper part of the back wall, illuminated by the third window. Below this is a black rectangular shape, in front of the wall. Here, after the teasing disappointment of the mirror in *The Music Lesson*, we have at last been granted what I believe is a distant glimpse of Vermeer's camera obscura. This tiny indistinct dark square is as much as he is prepared to reveal of his optical technique. As in *The Music Lesson*, no sign of the man himself: perhaps he is inside the booth.

How might this possible third window fit into the architecture of the room as a whole? (The question might seem like a digression; but its significance for Vermeer's camera technique will become clear.) Suppose that we assume it is a window of identical design—two casements and two fixed lights above—and of identical dimensions to those other two windows that can be seen in *The Music Lesson* and *The Glass of Wine*. Taking account of the known widths of these windows, the width of the pier between them, and the known dimension for the complete length of the room, we can combine all these elements neatly into the arrangement of Figure 47. Referring back to that drawing, see how these dimensions mean that the three windows can fit in a symmetrical arrangement, such that the (hypothetical) left-hand window ends flush with the back wall, just as the (visible) right-hand window ends flush with the far wall. Meanwhile the two piers between the windows are equal in width.

It transpires that the viewpoints of the six relevant paintings all fall more or less in line with the central vertical member of the frame of this third window (compare Figure 49). If the screen forming the front of Vermeer's camera met the wall at the centre of the window embrasure, this means that Vermeer would have had one casement inside his cubicle; and, more important, he could have opened or closed the external shutter on this casement, to black out the booth or let in light as he chose. The screen might have gone right up to the ceiling, in which case the fixed light would have had to be blacked out too. Alternatively the cubicle might have had its own roof at the level of the tops of the casements. This is what is assumed in the arrangement suggested in Figure 52.

If the reflection in the ball in *Allegory of the Faith* does show us the back of the room as Vermeer had it arranged for his camera, then this fragment of evidence seems also to suggest a booth with its own ceiling. This would explain why we see only the fixed lights of the third window and a brightly-lit area of back wall on a level with these lights. The sharp boundary of this illuminated patch of wall with the darkness below (see Figure 53a) would mark the top edge of Vermeer's cubicle.

The use of a camera obscura which completely encloses the artist implies, of course, that he works in near darkness. This has the benefit, as explained in Chapter 1, that the projected image appears relatively bright. On the other hand the artist cannot actually see the proper colours of his paints inside the dimly lit camera. If this *is* how Vermeer had his lens and booth set up, then we must assume that he used the camera obscura for two main purposes: to transcribe the projected image and to *study* compositional possibilities, optical effects, and relative tonal values.

A part of the underpainting itself might nevertheless have been carried out inside the darkened chamber. Chapter 2 illustrated how X-ray photography of *Girl with a Pearl Earring* has revealed a first stage of painting consisting of dark pigment applied to a light-coloured canvas, to produce a pattern which Gowing interprets as a more or less direct transcription of the observed tonal values of a camera image. A recent study of Vermeer's technique by Melanie Gifford seems to confirm this hypothesis, and for other pictures besides *Girl with a Pearl Earring*.[8] Vermeer appears to have begun these works with a brown or dark-coloured sketch, corresponding just to the disposition of shaded areas in the finished picture. He might have been able to apply this underpainting in dim light. As Gifford says of *Woman Holding a Balance*: 'broader areas of brown paint represented the masses of shadow, with the light buff color of the ground serving as the lights.'[9] The final paint layers conform closely to this first sketch in their boundaries between bright and dark.

For me, one of the most remarkable features of Vermeer's technique, as revealed by examination of the paint structure, is the fact that analysis has shown *no trace of black or dark-coloured line drawing*.[10] (There are the dark areas of tone, but no dark outlines.) This raises questions about how, precisely, the extreme precision of the perspective construction—as demonstrated in the last chapter—was translated into paint. In the work of the Delft architectural painters, by contrast, very detailed perspective outlines in dark chalk have been discovered by reflectography, under the final paint layers.[11] It is possible that Vermeer drew similar outlines in *light-coloured* chalk, traced from the camera, which would not show up under X-ray or infra-red analysis. But this is evidently not how he began working on the prepared ground, since the dark underpaint would then have covered any such chalk lines. (In this respect he cannot have followed the apparent method of the painter in his *Allegory of Painting*, who has drawn white outlines on a pale ground.) One might speculate that Vermeer was able to use his camera in different ways at different stages. At first he transcribed just broad areas of dark tone in monochrome. Later perhaps he made a more detailed tracing of the outlines of pieces of furniture, floor tiles, windows, and so on, and superimposed this in some way over the tonal background.

He may then, at a still later stage, have applied colours to the canvas outside the cubicle. He might have moved back and forth between the camera and the real scene out in the light, in order to remind himself of tonal contrasts and colour effects as they appeared in the optical image. An alternative is that he could have painted on the canvas *inside* the booth, with the window shutter open to give light, and from time to time closed the shutter to study the camera image.

Suppose then that Vermeer was using the camera image to transcribe areas of tone or to trace outlines. His optical set-up, as imagined here, has the virtue of simplicity, but it has the disadvantage that the projected image on the back wall, as seen from inside the cubicle, is both upside-down and reversed left-to-right (compare Figure 2a). The fact of the image being upside down would have been inconvenient for the purposes of tracing, but nothing worse. The fact of its being laterally reversed would have presented more serious problems.

There is one possibility which seems immediately attractive, but which we have sadly to discount: that Vermeer *did* simply transcribe the reversed image, and that the real room was not as we see it in Vermeer's pictures, but was itself mirrored in relation to the paintings, like the room Alice stepped into through the looking-glass. There is no objection to this idea as regards the architecture and some of the furniture. But the men and women in the pictures hold glasses and jugs, pluck stringed instruments, write letters, and paint paintings with their right hands. And, more conclusively, there are certain asymmetrical items—notably the maps and paintings on the far wall, and the virginals—that are evidently *not* mirrored.

It seems that Vermeer *was* faced with the task of reversing the projected image.[12] He could not, that is, have hung his canvas on the wall and traced onto it directly. There are several ways in which he could have achieved this reversal. Perhaps the most straightforward option is to assume that his camera was indeed of the simple room type as proposed in Figure 52, and that he did indeed trace the (reversed) image directly, but onto a piece of paper stretched on the back wall. He then pricked this tracing through with a pin, following the standard workshop technique for transferring working drawings or cartoons to canvases. He then turned his paper over, before placing it on the canvas and applying powdered chalk to transfer the design to the surface beneath. This technique of 'pouncing'—using powdered charcoal rather than chalk—was used extensively in the Delft pottery industry for applying standard designs to crockery and tiles.[13]

A second possibility is that the back wall—of which after all we see only a tiny area in the mirror in *The Music Lesson*—was not a blank wall without openings, but contained an internal window or door giving onto an adjacent room beyond. Interior windows are not unusual in Delft houses of the period, as de Hooch's and other contemporaries' paintings of domestic scenes demonstrate.[14] In this case we must imagine that Vermeer was not inside his camera when he worked on the image, but was on the other side of the opening, in the adjoining room.

He would have placed a screen of some translucent material, perhaps oiled paper supported on glass, in the opening. (Sashes or frames stretched with paper, linen, or silk were sometimes fitted inside glazed windows in the 17th century to act as sun-blinds. The materials could be varnished or oiled to render them translucent, and were often painted with landscapes or decorative motifs.[15]) The projected image, seen from the far side, would still have been upside-down, but no longer reversed left-to-right. This is exactly how the image in a plate camera is viewed, on a ground-glass screen substituted temporarily for the photographic plate. It is essentially the arrangement of camera obscura illustrated by Athanasius Kircher in his *Ars Magna Lucis et Umbrae* (Figure 8). The problem of transferring any tracing to canvas would still have remained. This could have been achieved by pouncing—but without turning the tracing over—as before.

The first arrangement—tracing from inside the camera—has the merit of being simpler, although there is the drawback that the artist's hand might to an extent have got in the way of the light coming from the lens. The second option—tracing from an adjoining room—has the advantage that the actual projected image would not then have been laterally reversed. This would have allowed Vermeer to make direct comparison of his painting in progress with the camera image itself (both of them upside-down).[16] On the other hand it is incompatible with the idea that the room seen in the paintings occupied the complete width of Mechelen or of Maria Thins's house.

Could Vermeer's camera itself not have incorporated some optical arrangement for rectifying the image? In Chapter 1 we saw two ways of doing this, both discussed in print before the mid-17th century: by means of a plane mirror set at an angle or by using two lenses in combination. The possibility of using a mirror was first mentioned by Benedetti in the late 16th century—although the idea does not seem to appear in any published illustrations of cameras until Johann Zahn's *Oculus Artificialis* of 1685–6.[17] It was Kepler who suggested using a pair of convex lenses to correct the camera image, and Christopher Scheiner who first illustrated a camera on this plan in 1630 (Figure 5). Zahn described the same principle, and he too showed it in a diagram (Figure 54).[18]

From a technical point of view the mirror arrangement seems feasible—although light is lost in any mirror reflection, and the resulting image therefore becomes fainter. The main objection for the present argument is that any geometrical reason for the coincidence of the six 'projected images' on the back wall then disappears. In a camera with a 45 degree mirror, the images would have to be projected onto a *horizontal* surface. As for a combination of two lenses, this is perhaps getting over-ingenious. On the criterion of Occam's razor, the simple type of single-lens camera seems preferable. Otherwise there is little evidence on which to decide between the various options. My main purpose in going into this topic has been to demonstrate that the problem of manipulating a laterally reversed image is far from insuperable.

Suppose that one of Vermeer's main uses for the camera obscura was to obtain precise outlines for the various shapes in a composition. It follows as a possibility that, when he came to paint, he might in some details have stayed with the outline but not represented exactly what he saw in terms of colour or texture. This is naturally a suggestion that it is impossible to prove, since in the end we have only the evidence of the paintings to tell us about the reality which *was* in front of him. But it does seem possible that we might have here an explanation for the phenomenon of the varying patterns of floor tiles, as well as for some other odd features that emerge in the perspective reconstructions.

We saw in Chapter 4 how there are three distinct patterns of marble floor tiles: separate white tiles framed in a lattice of black strips, white crosses on a black ground, and black crosses on a white ground. Even this third pattern is

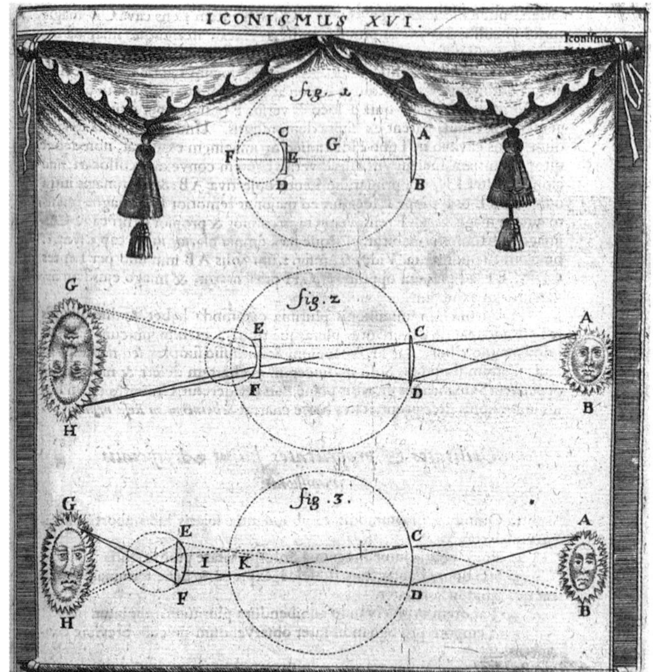

Figure 54
Diagram from Johann Zahn, *Oculus Artificialis* (1685–6) showing (below) a combination of two convex lenses to produce an upright image in a camera obscura.

not, however, precisely the same throughout, in that it meets the left-hand and far walls at different points in the repeat, in different paintings (Figure 55). Sometimes there is a row of white half-tiles along one or other of the walls, sometimes a row of black half-tiles.[19]

What does nevertheless remain the same in every picture, including *The Music Lesson* and *Allegory of the Faith*, is the underlying *grid*. Swillens was quoted earlier as suggesting—without any thought of Vermeer using the camera obscura—that in the matter of the tiles 'the painter allowed himself some licence'. Perhaps it is true that Vermeer traced the gridded outline of a single, real tiled floor, but then filled this standard grid with images of different patterns of black and white tiles?

What is more, if the smaller ceramic tiles in the two paintings in question really are half the dimensions of the marble tiles, as proposed in the last chapter, then the grids for the two types of tiles coincide exactly. The edges of the marble tiles lie on every alternate line of the grid marking the edges of

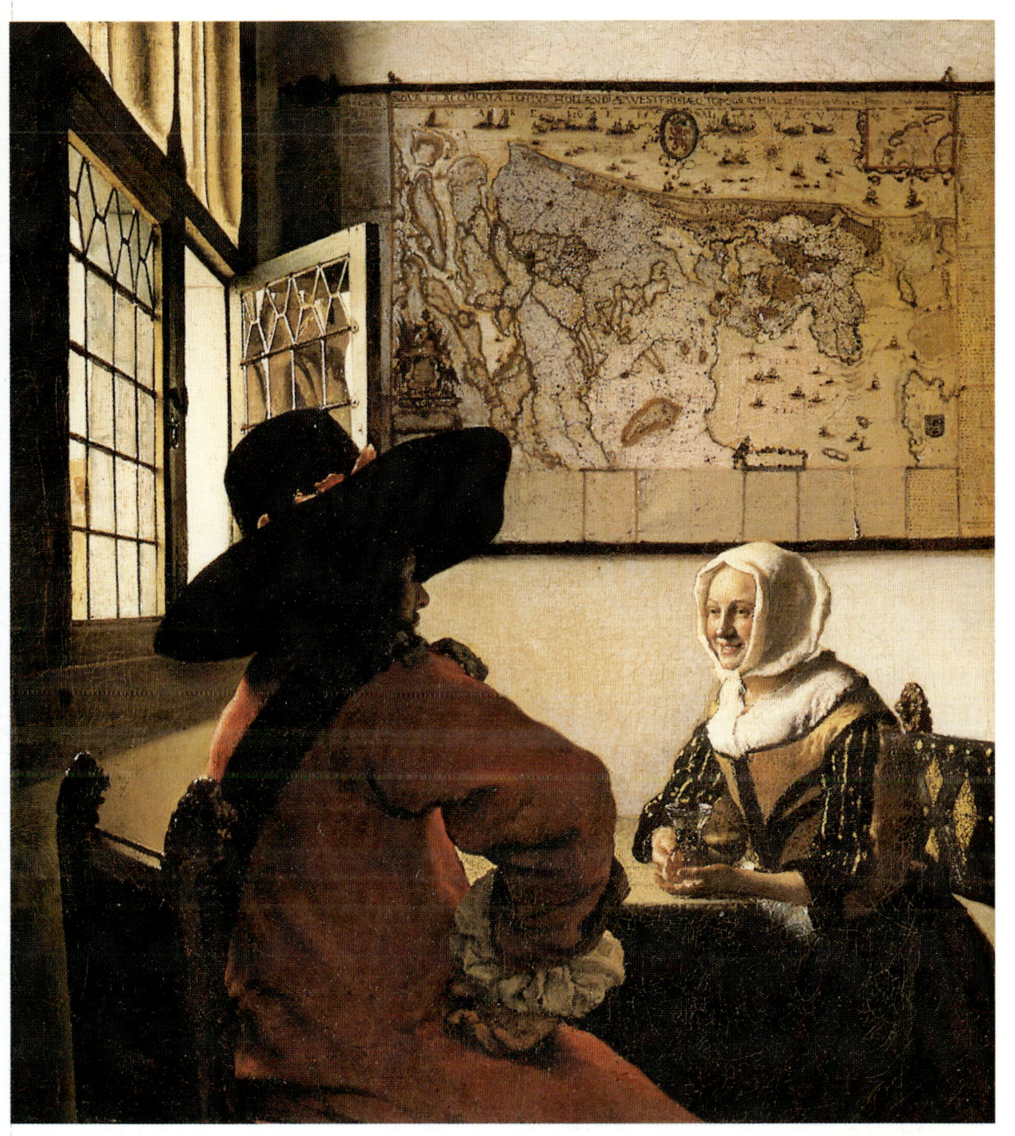

Vermeer, *Officer and Laughing Girl*, *c*.1658

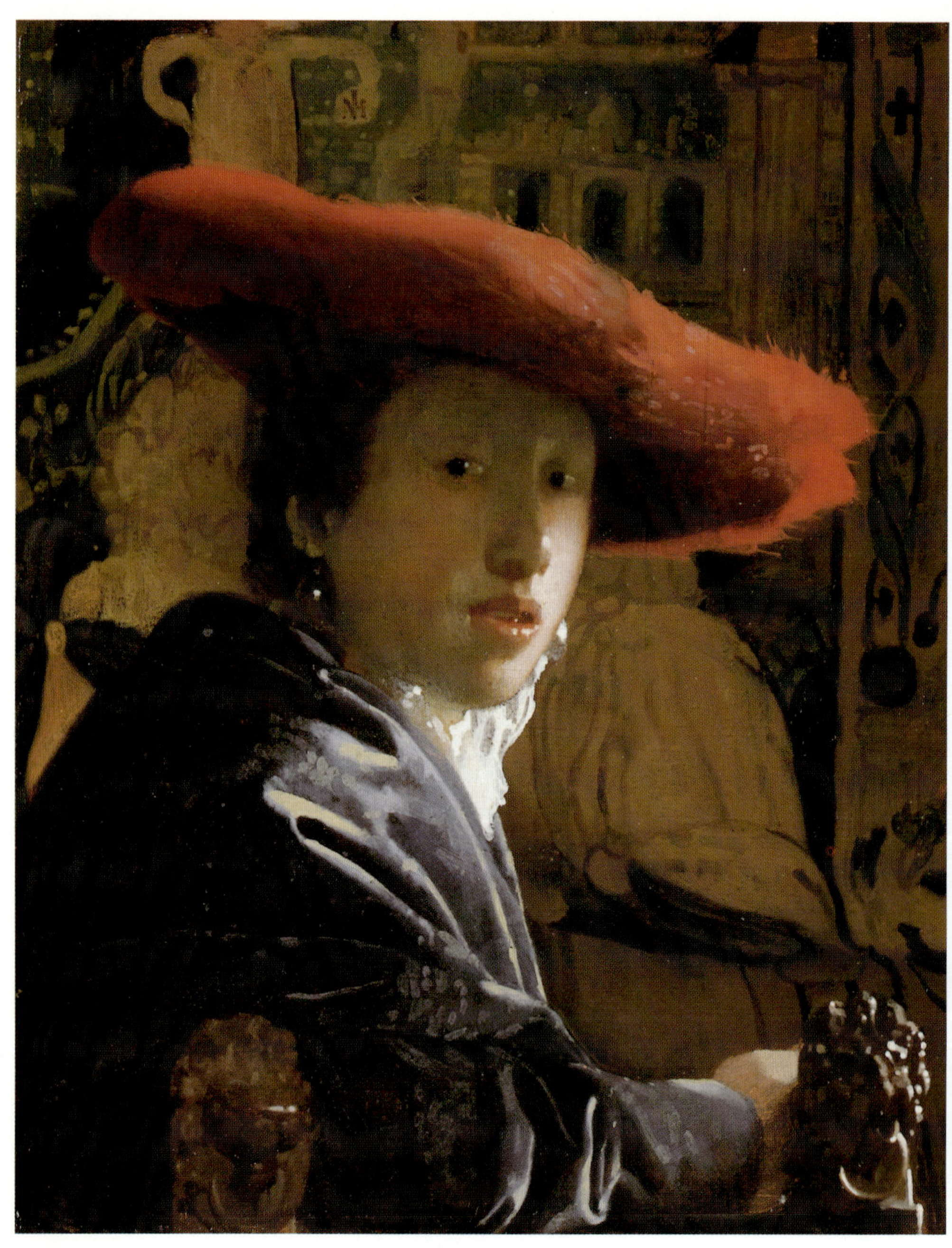

Vermeer, *Girl with a Red Hat*, *c.*1660–1

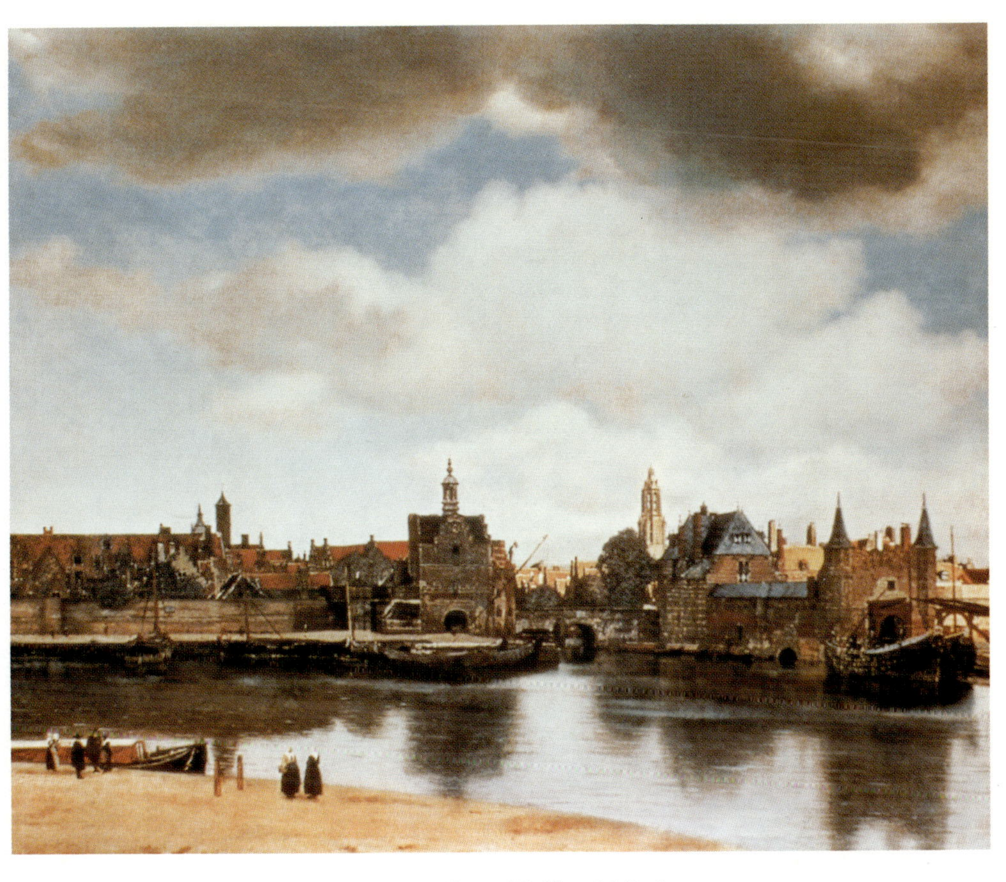

Vermeer, *View of Delft*, *c.*1660–1

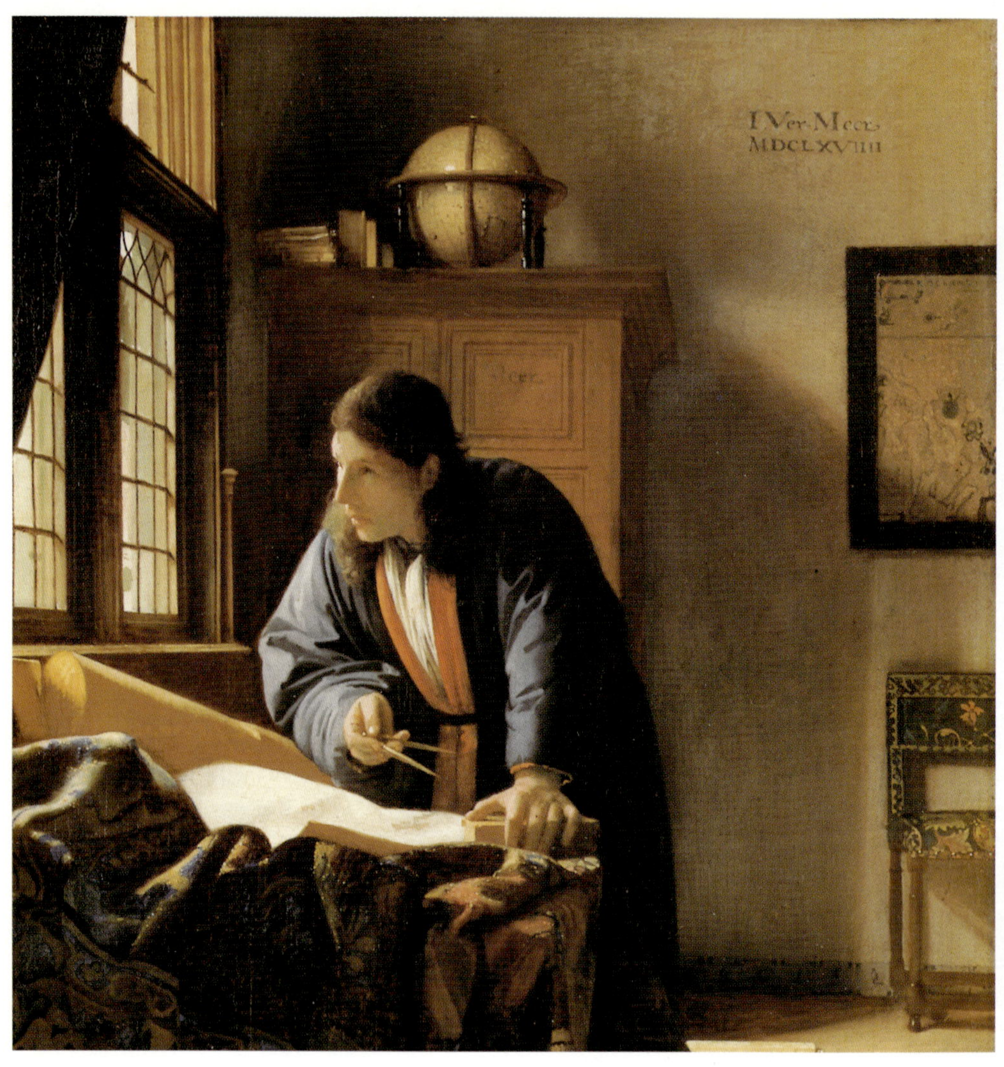

Vermeer, *The Geographer*, *c*.1668–9

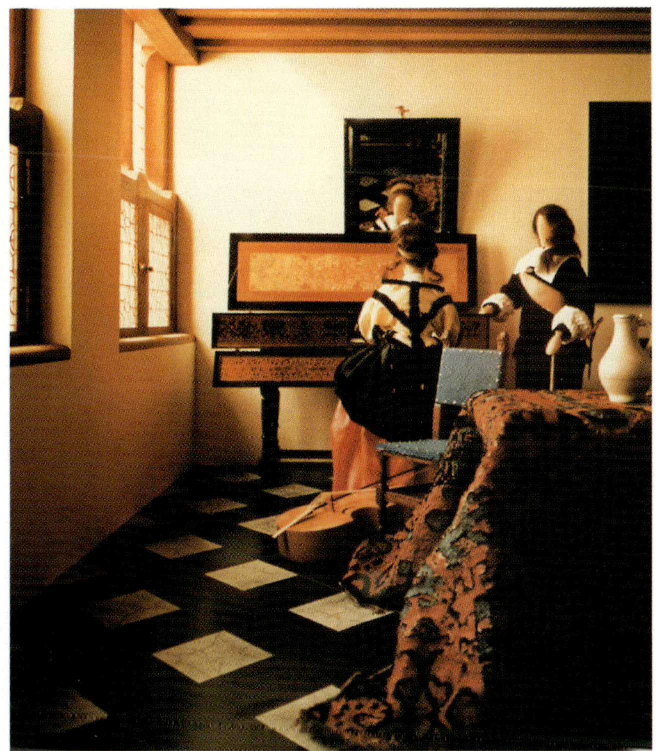

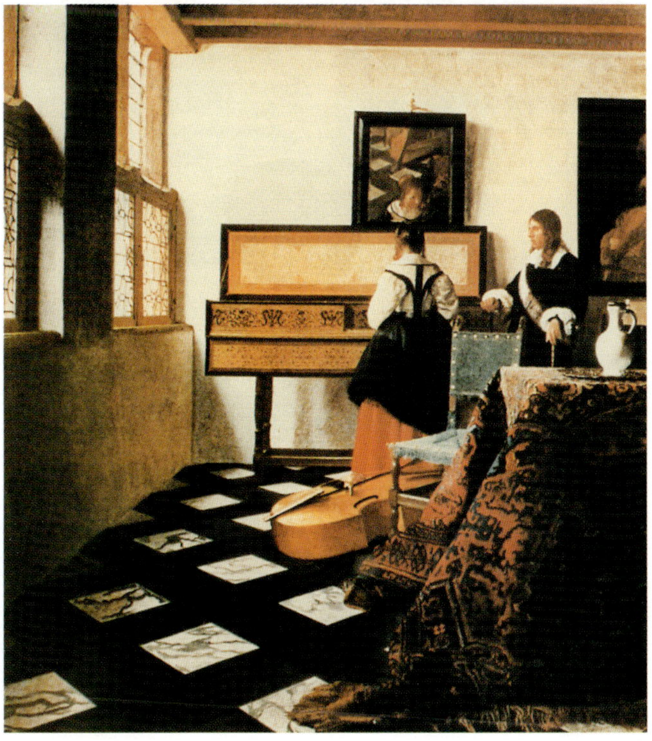

Photographic reconstruction of *The Music Lesson* (top), compared with the real painting (bottom)

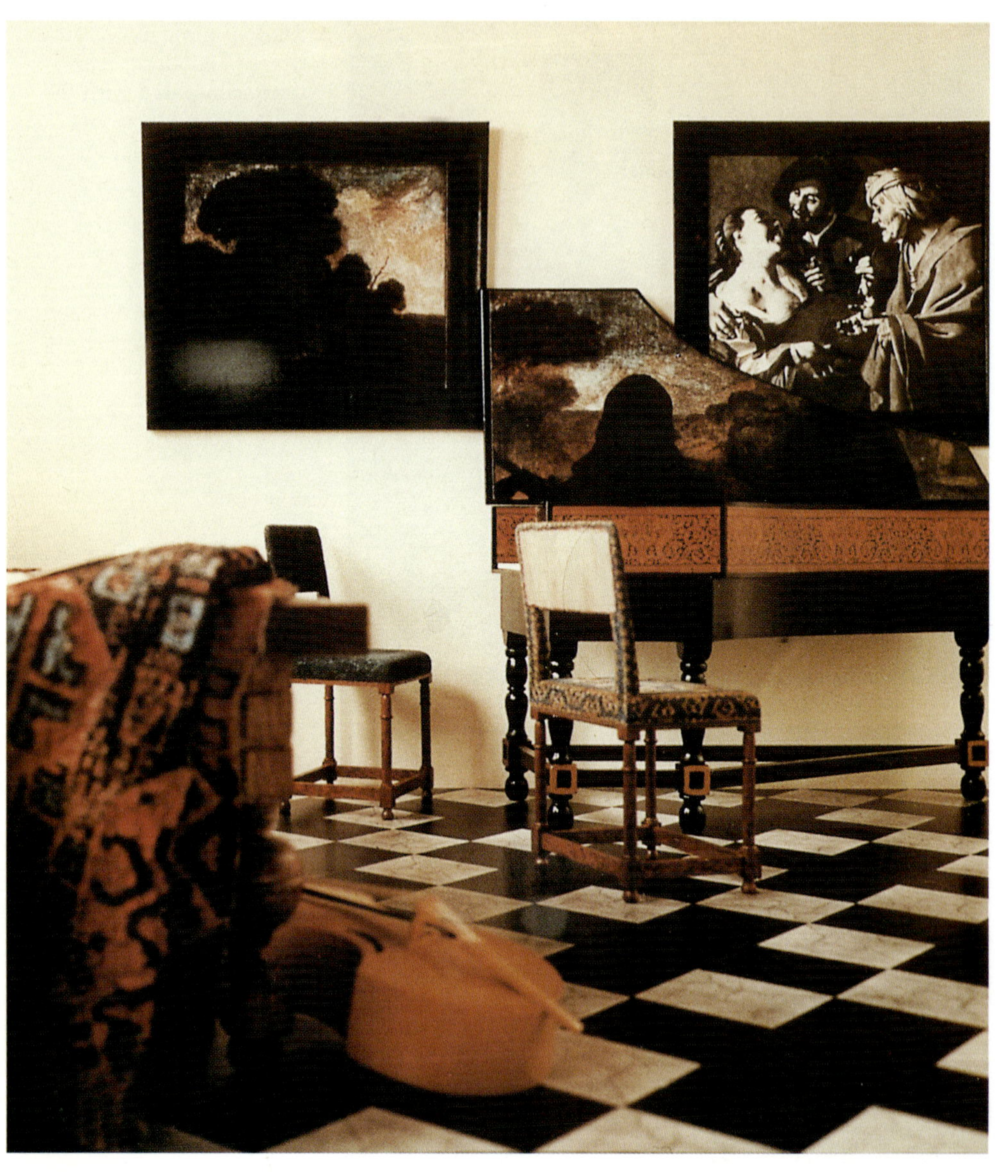

Photographic reconstruction of *The Concert* (left), compared with the real painting (right)

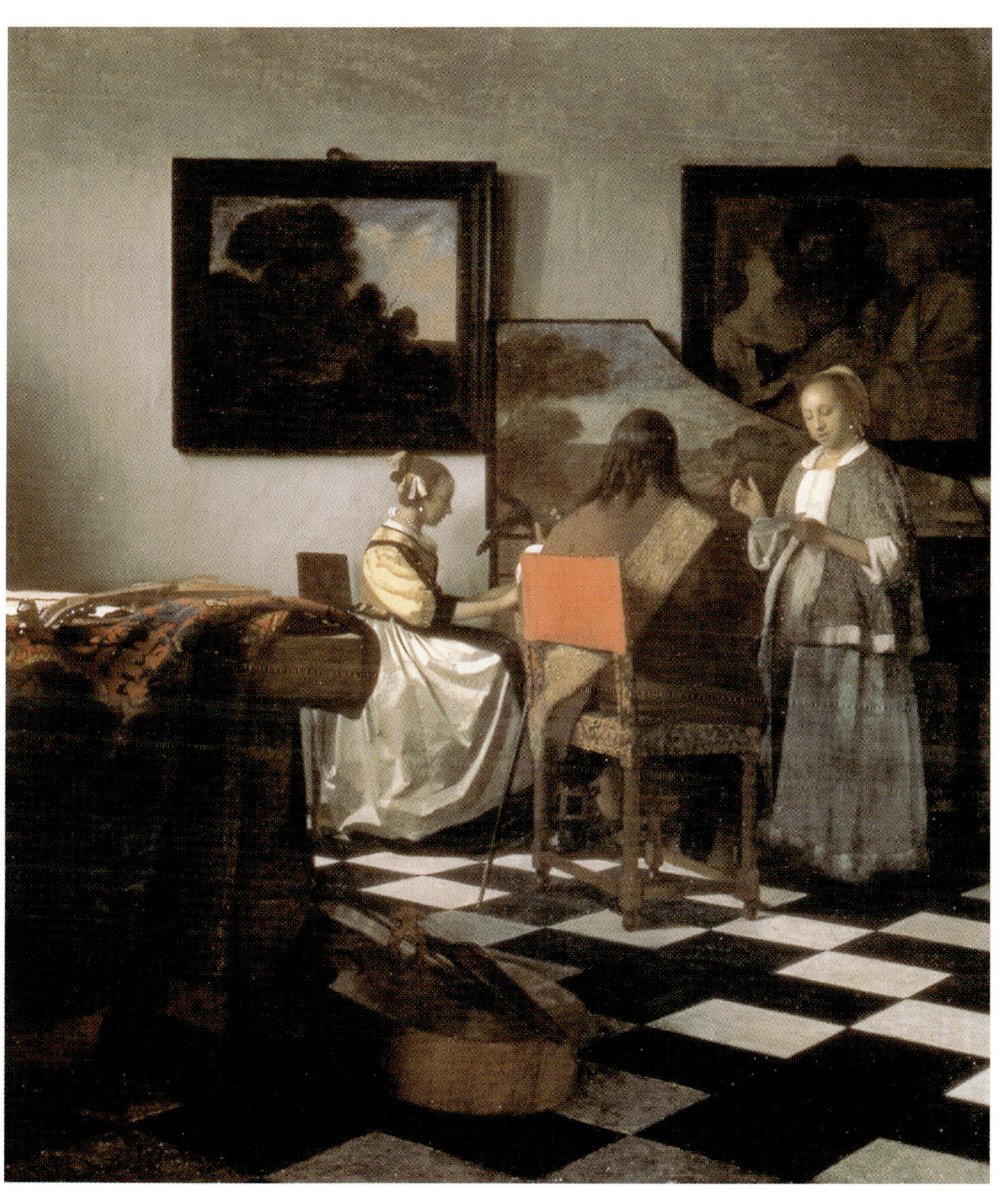

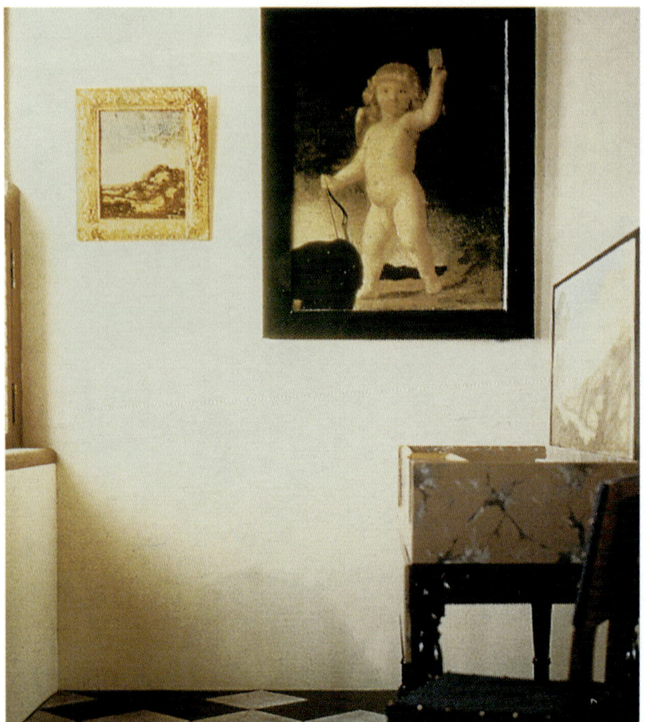

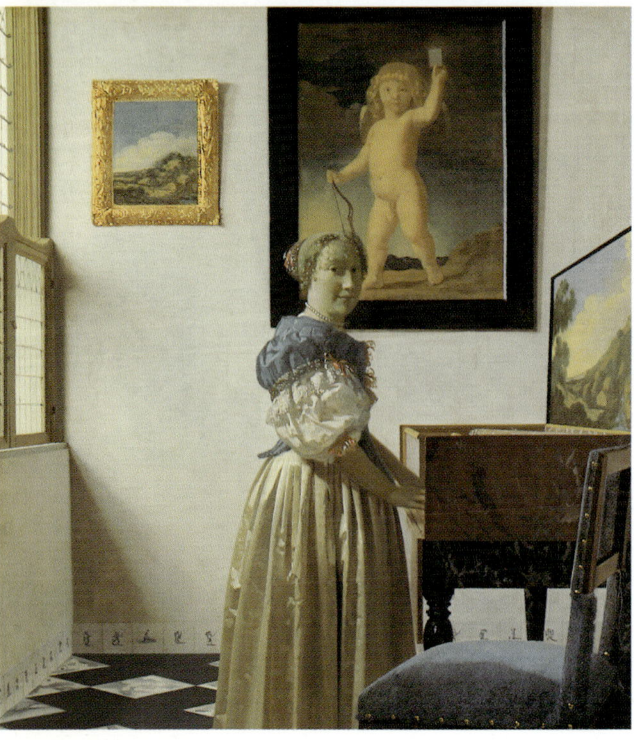

Photographic reconstruction of *Lady Standing at the Virginals* (top), compared with the real painting (bottom)

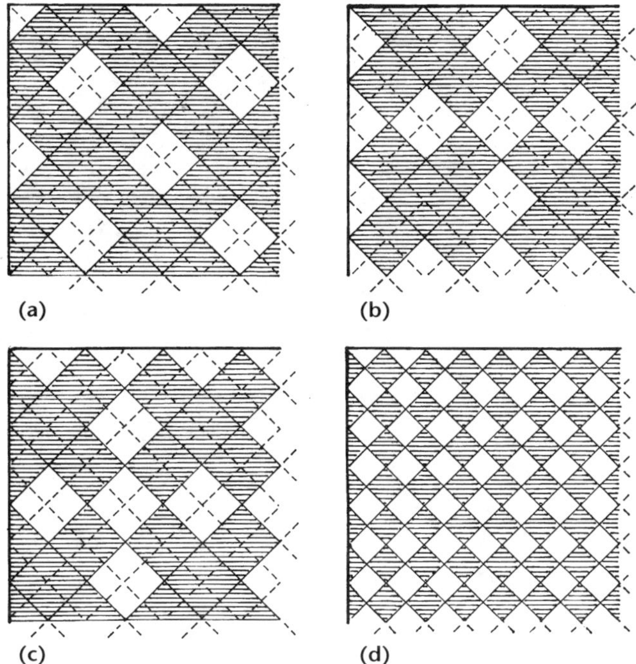

Figure 55
Grids of floor tiles in five paintings: (a) *The Music Lesson*, (b) *Lady Standing at the Virginals*, (c) *Lady Writing a Letter, with her Maid*, (d) *The Glass of Wine* and *The Girl with a Wineglass*. The grids of the larger tiles in (a), (b), and (c) all coincide with the grid of smaller tiles in (d)—shown in broken lines.

the ceramic tiles (see Figure 55). The possibility emerges that the real floor of the room was the ceramic tiled surface. Did Vermeer tire of painting this, after he had rendered it twice in the two early pictures *The Glass of Wine* and *The Girl with the Wineglass*? Did he decide to simplify his task from then on, as well as give his room a certain grandeur, by redecorating in (imaginary) marble?

A second instance where it seems that Vermeer might have traced an outline and then filled in with details which did not correspond to the real object, is in the instrument in *Lady Seated at the Virginals* and *Lady Standing at the Virginals*. This certainly appears to be the very same instrument in both pictures. The case has the same decoration of painted marbling on a light brown ground, and the stand is marbled in grey on black. The details of legs and mouldings are identical. The reconstructions in the last chapter gave overall dimensions for the instrument from the two pictures that were reasonably close—at least as regards height and depth.

The two painted lids are however entirely different. Both are landscapes, but *Lady Standing* . . . shows a mountain scene with buildings, in a black frame,

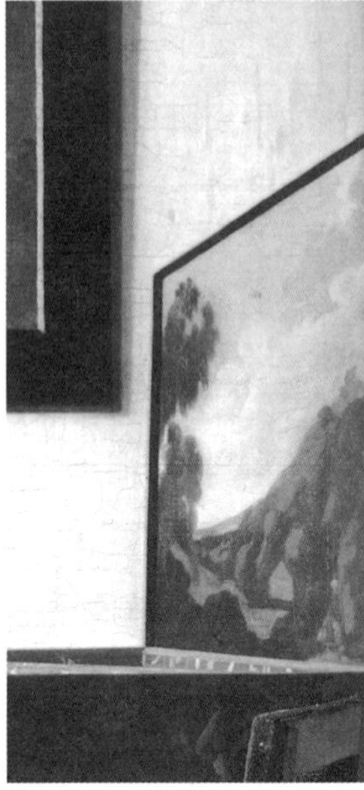

'Lady Standing ...'

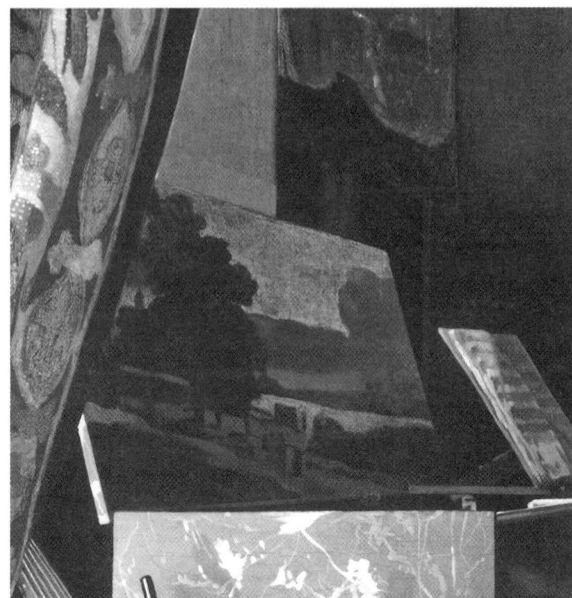

'Lady Seated at the Virginals'

Figure 56 Details of *Lady Standing at the Virginals* and *Lady Seated at the Virginals*, showing the painted lids of the virginals.

Lady Seated . . . a river bank with figures (Figure 56). The scene in *Lady Standing* . . . echoes the landscape painting on the far wall, which lacks the trees on the left, but has the same rising skyline, similar clouds, and the same two buildings at bottom right.

These landscapes on the lids appear quite normal as regards their perspective geometry. In truth they are most unusual. We see the lids in both instances at very shallow angles. If we reconstruct what the two landscapes would look like when seen frontally, we find that the scenes become quite excessively stretched out (Figure 57). These are *anamorphic* landscapes that only look realistic when seen very obliquely. They are like those puzzle pictures, often concealing politically subversive or pornographic subjects, which became popular from the 17th century, of which William Scrots's portrait of Edward VI and the skull in Holbein's *The Ambassadors* are two of the best-known examples.

Notice how the black frame around the lid in *Lady Standing* . . . is particularly misleading. Vermeer paints it at a standard width, both on the far edge and on

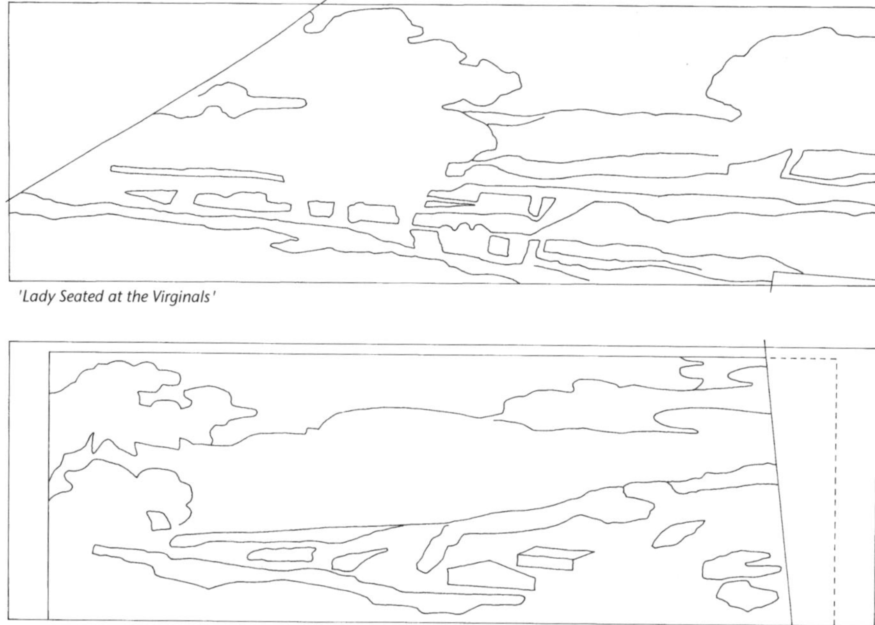

Figure 57 Reconstructions of the virginals lids in Figure 56, as they would be seen frontally.

the top edge. In order to project these apparent widths in the perspective view, however, the actual widths on the side and the top of the lid itself would have to differ very greatly. Vermeer has not allowed either for perspective diminution with distance in the frame along the top edge.

The explanation for these anamorphoses might be that Vermeer traced the outlines of the virginals in both cases, or studied their camera images, and found that the appearance of the actual painted decoration on the lid—in reality perhaps unlike either of the landscapes—was quite disturbingly foreshortened and distorted. He therefore decided to fill in the quadrilateral within the outline of the lid in each case with a composition suited to the surface of his painting, ignoring the steep perspective of the real surface of the instrument itself. We are not visually disturbed by this mild deception—indeed no Vermeer scholar seems ever to have remarked on it.

7 More evidence, from rebuilding Vermeer's studio

With detailed measurements of the architecture and furniture, it is possible to rebuild Vermeer's room both in model form and at full size. This chapter describes a series of photographic experiments made with a reduced-scale model and the recreation of the full-scale room for a television film. The experiments offered a means of testing the findings about the perspective geometry of the paintings; and a way of trying out the cubicle type of arrangement for the camera obscura in practice.

The model was built at a scale of 1:6, which means for example that a standing male figure is about 30 cm (12″) tall. The scale is such that a large (photographic) plate camera can be used in place of Vermeer's booth. The lens of this camera can be placed at each painting's viewpoint, and the photographic plate set in the plane of the back wall. The arrangement of the room for each painting can be mocked up with model furniture, decorations, and scale human figures. A photographic version of the composition can then be made for comparison with the original. With a model that is sufficiently detailed and realistic, it becomes possible to go beyond the limitations of a purely geometrical reconstruction, as in the drawings of Chapter 5, to study questions of lighting, shadow, tone, and colour.

Figures 58 and 59 show views of the model with the plate camera in position and furniture and figures in place for the reconstruction of *The Music Lesson*. The casements can all be opened inwards (although they are closed here) and the lower halves of the windows shuttered externally to control the lighting, as Vermeer did. Figure 58 for example shows the shutters closed on the back window next to the camera. (Detailed working drawings for the model room

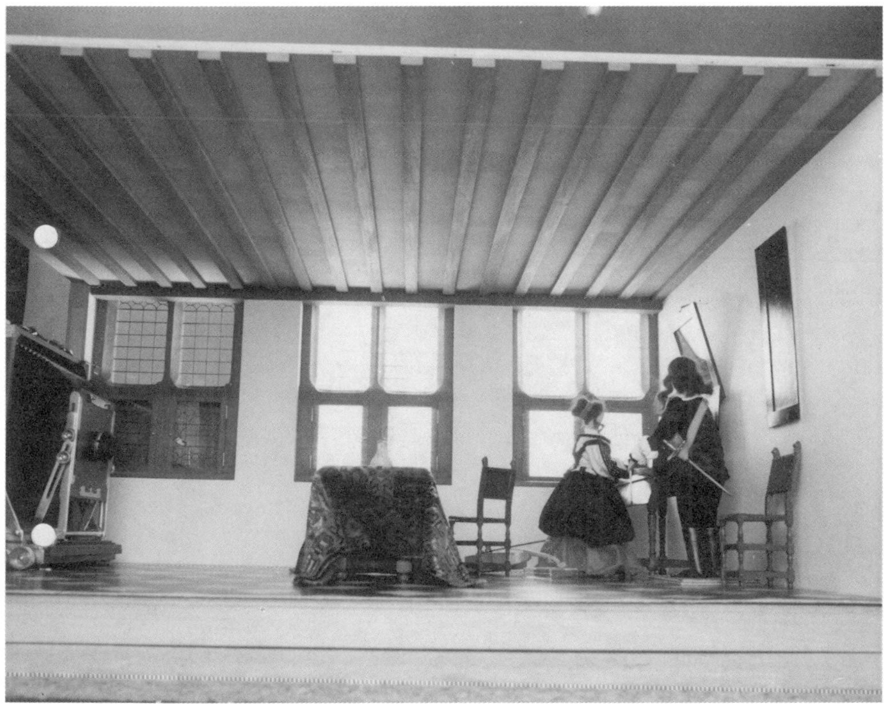

Figure 58 The model room from the side, with furniture and figures for *The Music Lesson*. The plate camera is at the left.

and furniture can be found on the website associated with this book, at www.vermeerscamera.co.uk.

The male and female figures for *The Music Lesson* were dressed in miniature costumes and given appropriate hairstyles, as the photographs show—although no attempt was made to simulate the details of faces and hands. (Articulated wooden lay figures of the kinds sold by artists' suppliers were used for the bodies.) In reconstructions of other paintings, however, the figures were omitted. This was partly because of the great difficulty of getting the clothes to hang properly, since it is impossible to scale down the weave and weight of cloth. But this is arguably not too serious an omission, since modelling the figures does little to test the geometry or lighting of the reconstructions. (Some shadows cast by the figures may of course be absent.)

The model room was lit with two lamps behind large fabric diffusers, as seen

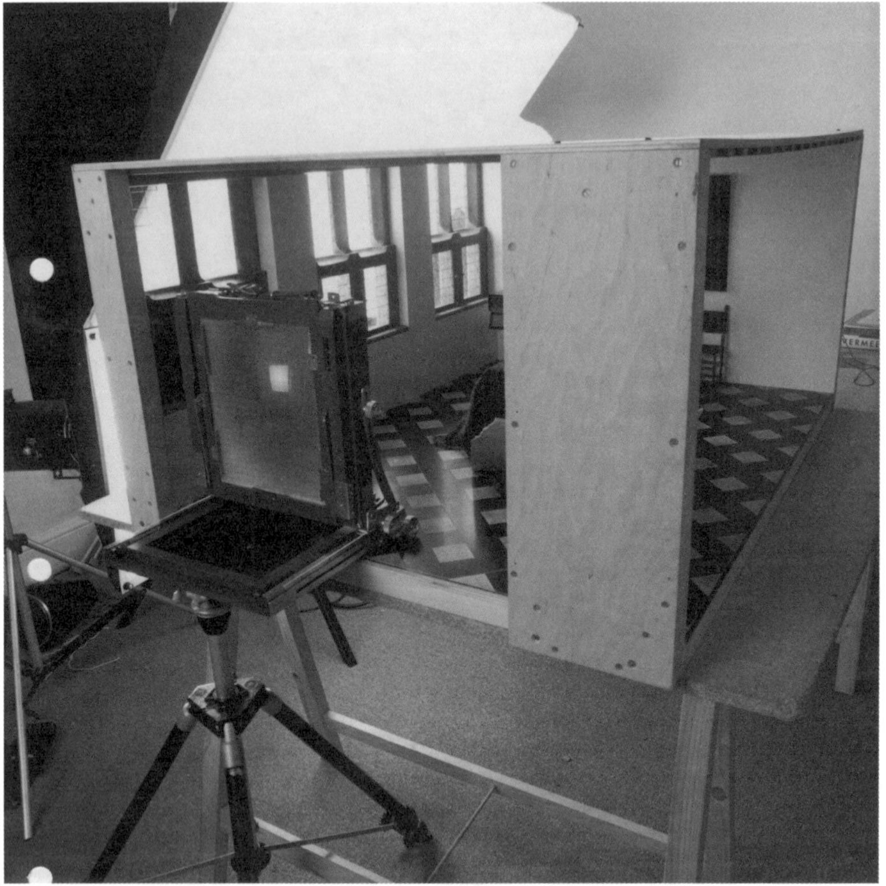

Figure 59 The model room seen through the aperture in the back wall. The plate of the camera is in the plane of this wall. The lighting is provided by two lamps covered with large fabric diffusers, to simulate north light.

in Figure 59. The notion was that this would approximate north light and avoid strongly directional illumination. This photograph also shows how the model room has no right-hand wall and has a large opening in the back wall. Some ambient light obviously enters from these sides too, although the principal illumination comes through the windows. (When some of the other paintings were reconstructed, the right-hand side and back were screened off.)

The working procedure for making each photographic reconstruction was as follows. The plate camera was moved into place, with the centre of the lens in the exact position calculated for the viewpoint as in the drawings in the last chapter. The ground-glass viewing screen on the back of the camera was then adjusted to lie in the plane of the interior face of the back wall of the room (see Figure 59). Furniture and figures were all set in position as calculated from the perspective reconstruction, using the grid of floor tiles as an additional guide to locating objects in plan.

The resulting image in the camera was then checked against the actual painting. This was done by making an exact one-sixth scale photographic reproduction of the painting and tracing the principal lines of the composition onto a clear transparent overlay. This tracing was taped (upside-down) onto the viewing screen of the camera in the appropriate position. Any discrepancies between image and tracing could thus be detected. This proved to be an effective means of adjusting such things as the folds of carpets or curtains, whose positions are very difficult to locate by geometrical construction. (The positions of items of furniture, previously calculated by geometrical means, were not of course altered just to fit the tracing, or the whole exercise would have become circular.) The results are illustrated below, alongside matching reproductions of the paintings themselves.

The Music Lesson

The Music Lesson (Plates 5 and 6) is perhaps the most successful of the reconstructions, in all aspects of geometry, lighting, tonality, and colour. When outlines traced from the real painting are superimposed onto the photograph, they show a very close match for all of the architecture of the room, including both the tiles and the details of the windows. The main discrepancy in the furniture is that the model virginals should be slightly longer, and should extend further to the right. The model jug, on the other hand, has been made somewhat too large. This has an odd effect on its apparent distance in the scene. (The costumes of the figures are also too wide, for the reasons already mentioned.)

Notice how the woman, table, and floor are correctly reflected in the mirror in the reconstruction, but how, at the top left of the mirror, where Vermeer shows his easel and stool, we see instead the wooden base and brass frame of the plate camera.

The gradations of light on the walls and the shapes and tones of the shadows are all quite accurately reproduced. Note in particular the shadows beneath the virginals and where the far window meets the far wall; the light on the sills; and the slightly brighter patch at the far end of the left-hand wall, just below the sill, where light is reflected back off the varnished virginals case. See too the gradient in brightness from left to right across the far wall, and the double shadows to the right of both the virginals lid and the mirror. All these effects were achieved without any great juggling of the photographic lamps, but simply by diffusing their light across the whole area outside the windows. The double shadows in particular were *not* a consequence of using two lamps, but arise because the light comes from two windows—the third window at the back of the room being too far away, and at too shallow an angle to the scene, to produce a well-defined third shadow.[1]

One anomaly, however, is that the double shadows cast by the mirror in the reconstruction make much wider angles with the frame than those in the painting.[2] The shadows of Vermeer's painted mirror are consistent with it hanging near to the vertical, as would be normal. In the model, however, the top of the mirror leans a considerable distance out from the wall, as the side view in Figure 39 shows. It must be at this angle in order for the relevant parts of the room to be reflected. Vermeer's rendering is therefore inconsistent. He wanted perhaps both to show a reflection of his own vantage-point by having the mirror lean forwards *and* to have the mirror appear to hang in a more normal, near-vertical position—requirements that are obviously not compatible in reality (although they are made to look so in the painting).

Figure 60 shows the whole of the coverage of the camera lens for *The Music Lesson*, including areas at the edges of the image which Vermeer excluded from his composition. In general Vermeer selects a squareish part of the whole image, quite close to the centre of the coverage of the lens—although usually slightly offset, as is the case here. The perspective distortion of the very nearest tiles is quite marked—Vermeer was clearly wise to exclude these.

The Concert

The reconstruction of *The Concert* (Plates 7 and 8) is a fair match to the original as regards geometry and colours, although the floor tiles and the far wall are the only visible features of the architecture of the room. The picture on the back

Figure 60 Entire coverage of the lens in the photographic reconstruction of *The Music Lesson*.

wall at the left is simply rephotographed from Vermeer's versions. The lid of the harpsichord is also reproduced photographically from the painting, which is why it has blacked-out silhouettes in the positions of the seated gentleman and the standing lady. The shadows on the back wall are, however, less successful than those in *The Music Lesson*. It is not possible in *The Concert* to determine exactly where the furniture is set against the back wall, relative to the left-hand wall (since the latter is not visible). It seems likely that, in the reconstruction, the scene has been set up too far to the right. Study of the painting shows a slanting shadow on the wall at the left-hand edge, just above the table-top, which might suggest that the window is very close. Figure 61 shows more of the coverage of the lens in the reconstruction. The shutter on the furthest casement has been closed and the result is a slanting shadow very similar to that in the painting. If the model furniture had all been placed further to the left, this shadow would have fallen in the appropriate position in the photograph. Van Baburen's *The Procuress* is reproduced here from a one-sixth scale photographic reproduction (in black and white) of the real work.

The gradation of light across the far wall in Vermeer's picture and the double shadows cast by the landscape painting, are both captured to a certain extent in the photographic reproduction, although not entirely accurately. The landscape does not hang out from the wall in the model at a steep enough angle. There is on the other hand some softening of focus in the viola da gamba, echoing Vermeer's own treatment.

Lady Standing at the Virginals

The geometry, colour, and tonal values of this picture are again well reproduced (Plates 9 and 10), the main geometrical discrepancy being in the window. As remarked in Chapter 4, the details of the leading of the casement and of the mouldings of the frame of the upper fixed light are peculiar to this painting. Otherwise, however, the window follows the general design of the windows in such pictures as *The Music Lesson*: hence it broadly resembles those in the model. However, there is a further discrepancy here between the painting and the reconstruction, which is that the top of the casement frame is appreciably lower in the painting—below the bottom of the gold frame of the landscape. The painting of the standing Cupid is slightly too small in the reconstruction,

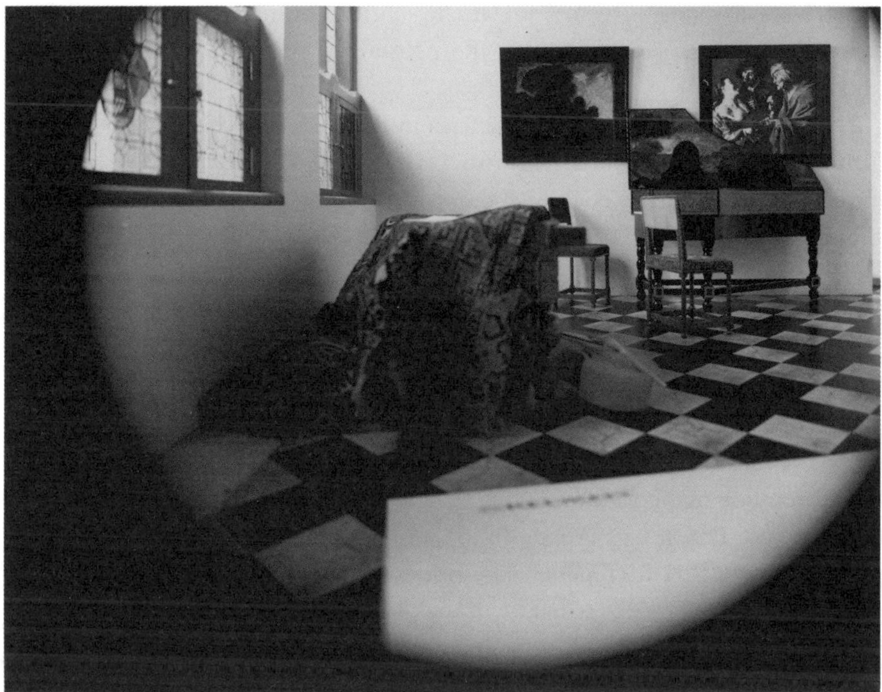

Figure 61 The greater part of the coverage of the lens in the photographic reconstruction of *The Concert*. Van Baburen's *The Procuress* is modelled with a reproduction (in black and white) of the real painting (compare Figure 45).

and the horizontal rail of the base of the virginals is slightly too low; but these are faults in the model-making. The colours of the anamorphic landscape on the lid of the virginals also appear very bleached in the photograph. This is because the model was made to match Vermeer's pale colours, but when the instrument is placed in its correct position the lid faces the windows and is very brightly lit. It should have been painted in stronger hues.

The Glass of Wine *and* The Girl with a Wineglass

Reconstructions of three more pictures—the remaining three whose 'projected images' are equal to the sizes of the paintings themselves—were less successful as regards lighting and colour, but still served to test aspects of the geometry.

These were all made using a 5″ × 4″ (half plate) rather than a 10″ × 8″ (full plate) camera, so that the entire area of the painting is not always covered in the photograph.

There are several minor discrepancies in the sizes and details of the model furniture in *The Glass of Wine* (Figure 62). The table is slightly too high, the jug is too large as before, the back of the lion's head chair bends too far back and should be straighter, the painting on the far wall has been made somewhat bigger than it should be. However, the major geometrical difference between photograph and painting is in the casements, which are too large in the photograph. The height to the top of their frames is also too great. The same is true for *The Girl with a Wineglass* (Figure 63). The measurements made from the perspective reconstructions showed a significant difference in these dimensions of window height between these two paintings and *The Music Lesson*. The model room is made to dimensions that conform more closely in this respect to *The Music Lesson* (see Appendix B). But this is perhaps not the complete explanation for this discrepancy in window sizes.

In both these pictures, the nearer of the two casements in the central window is open. This casement, with its stained glass coat of arms, is very close to the

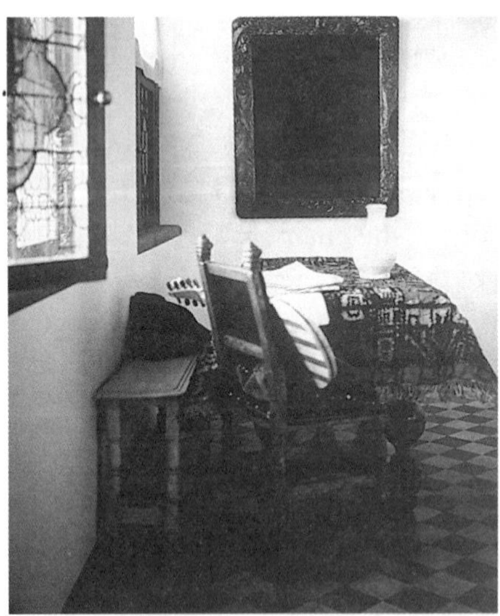

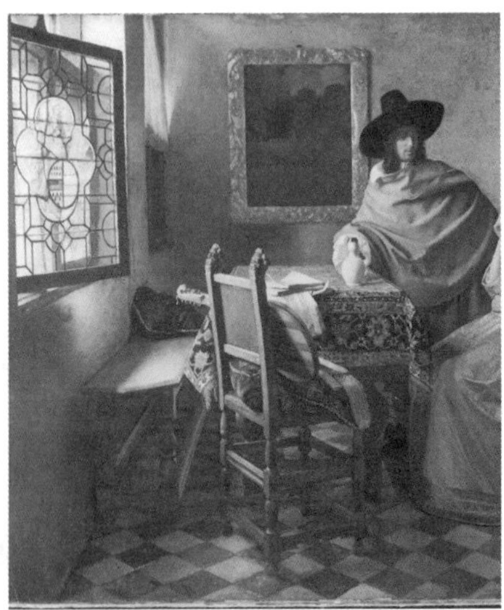

Figure 62 Photographic reconstruction of part of *The Glass of Wine* (left), compared with the same part of the real painting (right), Staatliche Museen Preussischer Kulturbesitz, Gemäldegalerie, Berlin-Dahlem. Oil on canvas 65 × 77 cm.

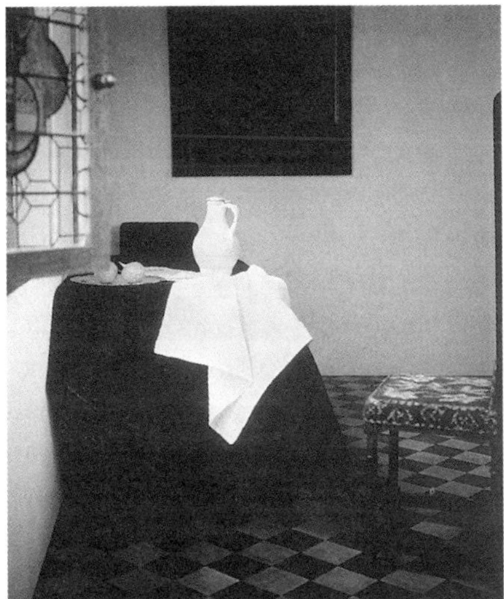

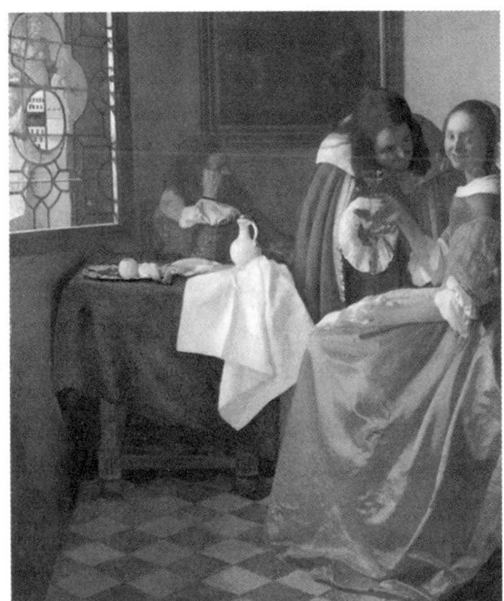

Figure 63 Photographic reconstruction of part of *The Girl with a Wineglass* (left), compared with the real painting (right), Herzog Anton Ulrich-Museum, Brunswick. Oil on canvas 78 × 67 cm.

picture's viewpoint in both cases, and as a consequence is bound to appear very large in a true perspective image.[3] Nevertheless, comparing painting with photograph for *The Glass of Wine*, we see that Vermeer has managed to include the whole of the casement, where in the photograph perhaps a quarter of its height disappears off the top, and almost half the width is lost at the left. What is more, the casement has been opened to a narrower angle in the model—and so is somewhat more foreshortened laterally—in order that its right-hand frame might lie in the correct alignment relative to the pier and window beyond. In *The Girl with a Wineglass* the open casement is again too tall. (There is no sign in this painting of any second farther window, but this could be because the background has been substantially repainted, as previously mentioned, or, of course, because the artist has decided to leave it out.)

We are stepping on to dangerous ground here—but is it possible that Vermeer has not, in these two pictures, represented this near window exactly as it appeared in the projected image, since had he done so it would have appeared alarmingly oversized? Are they in this respect *collages* or composites, in which the window wall and casements have been scaled down relative to the rest of the scene, or have been traced from an image projected at a greater distance, further behind the line of the back wall of the room? I have already remarked on the fact that, among Vermeer's interiors, these two paintings are usually given early dates. Perhaps the artist experimented in these two instances with viewpoints that were close to the left-hand wall (compare Figure 49), and which thereby brought the nearer casement into view. Later, realizing the problems involved, he moved his viewpoints further to the right.

There are some perplexing features too in the *lighting* of *The Glass of Wine*, which the reconstruction wholly fails to reproduce or explain. The two casements in the far window are obviously shuttered. But why does the diagonal shadow on the far wall strike *upwards* from the top of their frame? Why is the bench below the near window so sharply divided across its width into light and shade? And why is there such a brightly-lit patch on the wall below the sill of this window? I have no answers to these questions.

Lady Writing a Letter, with her Maid

This picture presented problems with lighting in the photographic reconstruction (Figure 64), since much of the scene is in shadow, and it proved difficult to curtain the window wall exactly as Vermeer shows it. (The shutter to the casement behind the white curtain is closed.) The match to the geometry is nevertheless accurate in all details. Incidentally although this picture, like some others (*Allegory of the Faith, Lady Seated at the Virginals*), shows the room in a dim light, it would not have been essential for Vermeer to preserve this lighting at all times while tracing images in the camera obscura. He could have pulled back the curtains and opened the shutters to see outlines and positions more clearly in the camera image, and replaced them again when studying tonal values or painting directly from the scene.

To recapitulate: these six photographic reconstructions have all been made with the plate of the camera exactly aligned with the back wall of the model room. Despite the discrepancies in the sizes—especially the heights—of the windows, and the speculations which these have given rise to, the most important finding is that the images of the floor tiles in the photographs match those in the paintings precisely in all cases. This is the critical result which confirms the correctness of the calculated viewpoints and determines the sizes

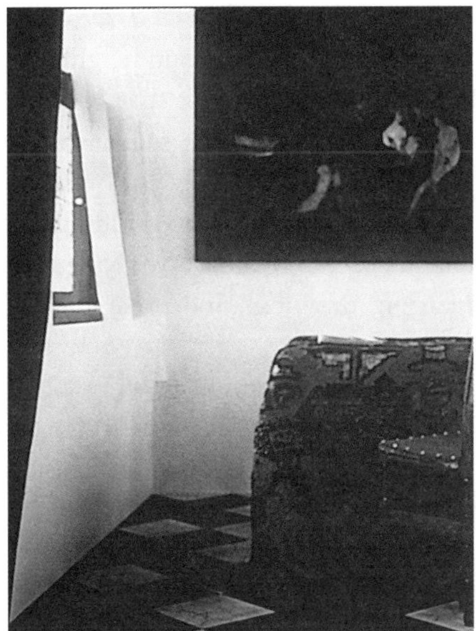
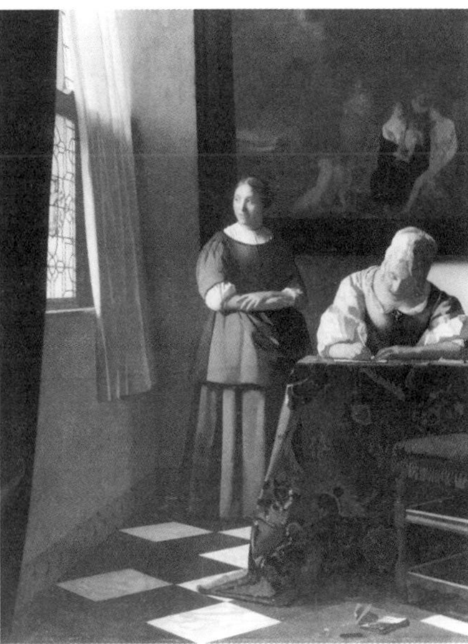

Figure 64 Photographic reconstruction of part of *Lady Writing a Letter, with her Maid* (left), compared with the real painting (right), National Gallery of Ireland, Dublin. Oil on canvas, 71.1 × 58.4 cm.

of the resulting projected images. The correspondence of these projected images with the sizes of the paintings themselves, revealed in the last chapter, has been amply confirmed. It should be emphasized that this whole photographic exercise provides, in effect, a second independent test of the calculations described in previous chapters.[4] If those were incorrect, then the photographs would not match Vermeer's originals geometrically in the way they do. It remains to re-examine briefly the five remaining paintings for which geometrical reconstructions were made, as described in Chapter 5. In none of these cases does the size of the 'projected image' at the back wall match that of the actual painting.

Allegory of Painting *and* Allegory of the Faith

Allegory of Painting and *Allegory of the Faith* share one outstanding geometrical property which distinguishes them from all of Vermeer's other interiors: they are much larger pictures. They measure 120 × 100 cm and 114 × 89 cm respectively: *The Music Lesson* and *The Concert* are only just over 70 cm in height and most of the others are less than 50 cm (Figure 65). Is it possible that Vermeer traced images for these two paintings at the back wall in the same way as described for other pictures, and then enlarged these drawings up by some means to the sizes of the canvases?[5] He might have done this by gridding the drawings up or indeed with the aid of a camera obscura, which as we have seen was used extensively for reducing or enlarging drawings and other images before the introduction of photography.

No attempt was made to recreate either of these compositions photographically. However, the model *was* used to examine the question of the reflection in the suspended glass ball in *Allegory of the Faith*, as discussed in the last chapter. The furniture and decorations were not mocked up in every detail, but the draped table and open Bible were modelled, and the seated woman was represented by a lay figure in a dress. The model room was completely blacked

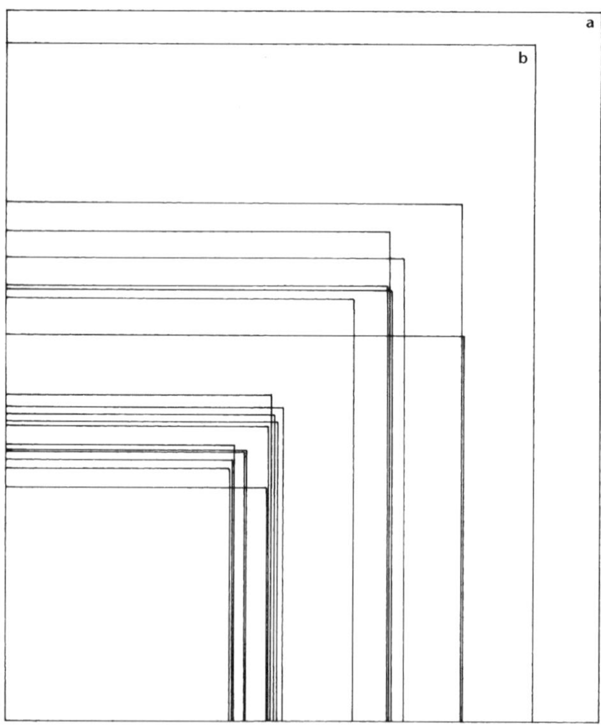

Figure 65
Comparative sizes of Vermeer's paintings of interiors (to scale). *Allegory of Painting* (a) and *Allegory of the Faith* (b) are much the largest.

out and the windows suitably shuttered and curtained. The ball itself was modelled with a silvered glass sphere of the correct diameter (see Figure 53b). This reproduced the pattern of the reflected windows with some success, as the figure shows. In the position where Vermeer's putative camera appears in the painting, we see the reflection of the (photographic) camera taking this picture. Notice how the two fixed lights at the back of the room seem very nearly to merge into a single bright rectangle, just as Vermeer shows them. (In the photograph, on the other hand, the reveals of the piers are strongly lit and appear as bright vertical strips, which are not seen in the painting.)

The Love Letter

As we have seen, it is dubious whether *The Love Letter* can be argued to show the 'same room' as the other paintings, since the only architectural element in common is the grid of floor tiles. Suppose however that it *is* the same. Then the viewpoint of the perspective falls some 1.4 metres behind the back wall. It is attractive to think that we might here be looking through the very doorway, in that wall, whose existence we have guessed at, as one arrangement for his camera obscura. It is the door, that is to say, in which Vermeer would otherwise have placed his translucent screen. But a beautiful theory is killed by the ugly fact that the estimated depth of the room seen in *The Love Letter*, from front to back, is considerably shorter than the depth of the room deduced from *The Music Lesson* (compare Figures 33 and 35). Indeed there are many details which, on closer examination, combine to suggest that the foreground parts of the composition in *The Love Letter* are assembled by Vermeer rather as framing elements for the view beyond, and not as a completely coherent representation of some real door opening.[6]

Lady Seated at the Virginals

Lady Seated at the Virginals presents an enigma. The viewpoint for this painting falls a short distance *behind* the back wall of the room. Thus its geometry is not compatible with a 'projected image' of the same size as the picture itself, as is the case with its pendant *Lady Standing at the Virginals*. Some critics have questioned whether Vermeer painted the two pictures as a pair. Certainly they have not always been in the same ownership and been hung together, as they are today in the National Gallery in London, but it is difficult to believe that they were not conceived in some degree as matching or complementary works. The canvases are virtually identical in size, and besides the common subject matter and the same virginals, the two ladies—or is it one and the same lady?—even wear the identical dress, with an added overskirt for *Lady Seated*. . . .

The proposal has often been made that *Lady Standing* . . . symbolizes pure love, the Cupid holding the playing card signifying fidelity to one alone, while *Lady Seated* . . . denotes profane love, as signalled by van Baburen's *The Procuress*. The darker tones of *Lady Seated* . . . add weight to this interpretation, as do the amply feminine curves of the viola, echoing those of van Baburen's whore. All questions of symbolic meaning aside, however, the fact of the paintings' dimensional and perspective similarities makes the disparity between their calculated viewpoints a little surprising.

A further oddity is that Vermeer has not shown van Baburen's *The Procuress* at either its true size or with its true proportions. Vermeer has both enlarged the painting and *stretched* it somewhat vertically. It might be that *Lady Seated at the Virginals* is a composite of tracings of parts of images onto planes at different distances and with different positions of the lens. Perhaps this painting was not after all conceived by Vermeer as a pendant to *Lady Standing* . . ., but was a variant on the same theme, painted later?

Woman with a Lute

The last picture to show some area of tiled floor, and whose space could thus be reconstructed, was *Woman with a Lute*. However, this area is very small—we can see perhaps only half a dozen tiles—and is confined to the right-hand edge of the picture. There are thus very few lines along the edges of the rows of tiles that can be used to find the 'distance points' of the perspective. The precise angles at which these lines lie are also difficult to measure.

What is more, such of the tile pattern as can be seen does not meet the far wall in a neat row of half-tiles as in all other paintings, but cuts the pattern at an intermediate position. *Woman with a Lute* is in poor condition and has suffered extensive repainting. It seems at least possible, therefore, that the line where floor meets wall has at some time been shifted in the restoration process.[7] If this is the case, then the effect of course would be to throw out any calculations of the distance of the viewpoint from the ends of the room.

The position of the viewpoint as calculated from the reconstruction discussed in Chapter 5 is slightly further forward than those of other relevant paintings. If it was assumed that the floor should properly extend somewhat further than the painting presently shows, then the viewpoint would move backwards, and it is *possible* that this picture too might then give a 'projected image' at its true size on the back wall. There are five more paintings of which this *could* also be true, but the matter cannot be decided because the floor is either not tiled or cannot be seen at all: *Girl Interrupted at her Music*, *Woman in Blue Reading a Letter*, *Woman Holding a Balance*, *Woman with a Pearl Necklace*, and *Young Woman with a Water Jug*.[8]

Reconstruction of the room at full size

In 1989 BBC Television built a full-size version of the room for a programme devoted to this work on Vermeer. They followed the measurements obtained from the reconstructions detailed here. The set was dressed as for *The Music Lesson*. The parts of the lady and gentleman were played by the programme's presenters. Figure 66 shows the set seen from the side, to give an idea of its construction. (Only that small part of the ceiling which is visible in the painting was actually built.)

The main point of interest here is that the set incorporated an actual camera obscura, of the cubicle type proposed in the last chapter. A simple convex lens of 10 cm diameter was used, comparable with those sold for school physics experiments. There was a door-sized opening in the back wall, at the appropriate position, with a translucent screen (made of plastic drawing film) stretched across it. This made it possible to film the projected image from the further side of the screen—from outside the set. Thus the arrangement was essentially that of Athanasius Kircher's camera. (A television camera positioned *inside* the cubicle, in order to film the image projected on a solid wall, would of course have blocked its own light.)

The result of using a lens with such a relatively large aperture was a bright image, at the size of the actual painting, quite clear and detailed enough to

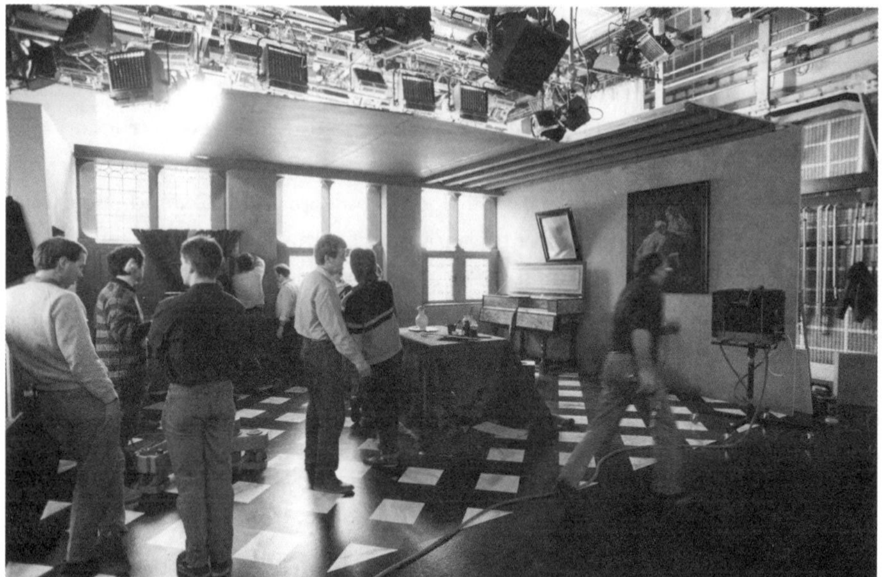

Figure 66 Set of the BBC Television reconstruction of the room at full size, dressed as for *The Music Lesson*.

trace, indeed quite bright enough to film. In the BBC set the lighting, although it came almost entirely from the windows as in Vermeer's room, was artificial and powerful. On the other hand the human eye is much more sensitive than the television camera. Thus it would have been possible for an artist to work from a less bright image than this experiment produced. Such a demonstration proves beyond question the feasibility of a room-sized camera, producing images of the required dimensions, and using modest optical technology. In the next chapter I take up the question of whether such technology could have been available to Vermeer.

8 Arguments against Vermeer's use of the camera

A number of writers on Vermeer have questioned the feasibility of a camera technique, or doubted the possibility that he might have copied camera images more or less in their entirety. Some of their arguments have been rehearsed already. In this chapter I take up the theme of objections and difficulties to the camera theory. I also examine alternative methods by which Vermeer might have obtained the perspective structure of his compositions, and discuss whether these are compatible with the geometrical findings of Chapter 6.

Jean-Luc Delsaute has expressed doubts as to whether the box type of camera obscura was of any use at all in the 17th century as a tool for artists.[1] He claims that cameras capable of being used in drawing from nature were introduced only in the early 18th century; that in the 17th century, cameras had 'virtually no practical applications'; and that where the 17th-century literature does refer to the camera in the context of picture-making, this is only to mention that painters might profit by *studying*—not transcribing—its images. Delsaute is prepared to allow that Vermeer might have known about the camera, and perhaps was interested in the images it could produce. But 'it seems rash', he says, 'to continue to believe that the camera obscura was one of the tools with which he worked.'[2]

In reaching this conclusion Delsaute relies heavily on Johann Zahn's *Oculus Artificialis*.[3] Published ten years after Vermeer's death, this optical work describes and illustrates various designs of small box-type camera (see Figure 41). Zahn calls these *cistulae parastaticae*, and Delsaute characterizes them as 'mobile camera obscuras with external observers'. Zahn also illustrates a different optical device, the *arcula delineatoria*, which is something like a modern light-box or Grant projector, intended for copying existing drawings or prints. Zahn goes on to describe how the *cistula parastatica* and the *arcula delineatoria* may be combined, to create a composite instrument with which it is possible to draw from nature. Delsaute interprets this passage as meaning that the simple box camera on its own *cannot* be so used.

Such an inference is quite untenable. Assuming a box-type camera is equipped, as it must be, with some sort of translucent screen, then the image projected onto that screen can readily be traced onto a transparent medium. The designs in Zahn's book all have such screens. Simple box cameras were used in this way by numerous artists, both professional and amateur, throughout the 18th and 19th centuries, as illustrated in Chapter 1. The instruments that Fox Talbot and the French pioneers converted into photographic cameras were, again, standard box types of camera obscura, sold for use by landscape artists. If this type of small camera was usable for drawing (without any additional *'arcula delineatoria')* after 1700, then clearly it could have been so used before 1700. Indeed, we do not have to rely on historical evidence here: simple practical experiments with existing box cameras show that drawing with them is perfectly feasible and unproblematic.

In any case, this particular point is of limited relevance for my own argument, since I have proposed that—with the possible exception of the two small Washington panels—Vermeer used not a box camera but a booth, in which he was completely enclosed. Delsaute refers to this kind of design as a 'camera obscura with an internal observer'. He reproduces an illustration from an early 18th-century instruction manual by W J 's Gravesande, showing a portable camera resembling a sedan chair, with a 45 degree mirror and convex lens in the ceiling (Figure 67)[4]. The principle is shown diagrammatically in Figure 6. He claims that this is a 'totally new model' which made the device into a practical tool for artists for the very first time. He says that 's Gravesande did not originate the design, but the implication is that this type did not exist before the 18th century.

It is not clear in what feature, precisely, Delsaute believes the novelty to lie. Presumably it is in the overhead lens and mirror which, as illustrated in Figure 6, can produce a correctly oriented image on a horizontal drawing surface. This arrangement may indeed have been introduced only in the late 17th century. Zahn shows the configuration (see the bottom of Figure 41), but on a box camera. I know of no earlier description or illustration than 's Gravesande's, of

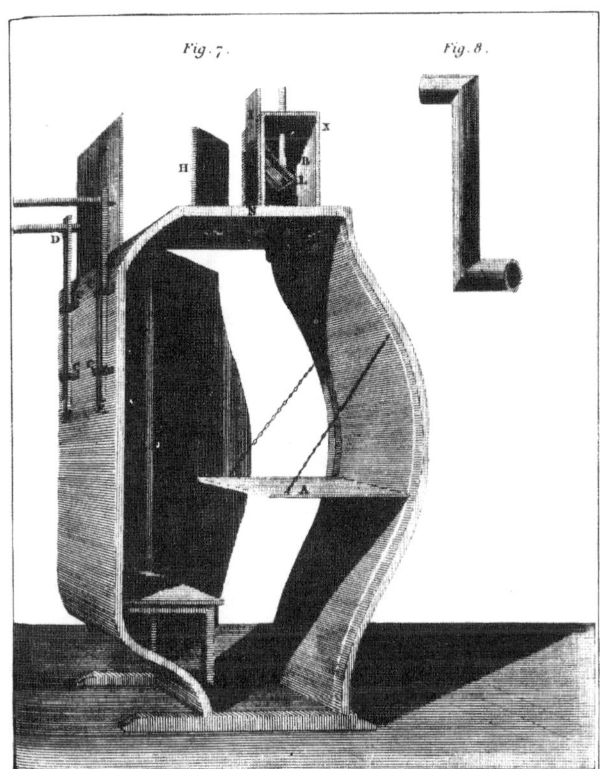

Figure 67
Portable camera obscura resembling a sedan chair, with overhead lens tube and mirror (in the box on the roof), from W J 's Gravesande, *Essai de Perspective*, 1717. A panoramic view can be made in stages with the aid of a second vertical mirror which turns on a vertical axis. The large vertical board on the roof is an optional extra, for mounting prints or paintings that are to be enlarged or reduced in the copying process.

a booth-type camera with vertical lens tube. The design is certainly an improvement over other types in which the image is inverted and/or reversed, but this development is hardly such as to make the difference between uselessness and usability. (It has the *dis*advantage that some light is lost in the mirror reflection.)

Otherwise, the principal characteristics of 's Gravesande's design are that the user is enclosed inside the camera and that the device is portable. Both features have their advantages, but they are certainly not new in the 18th century. Indeed, historically, the 'camera with an enclosed observer' precedes the 'camera with an external observer', as we have seen. Kepler's camera in a tent, as described by Wotton in the 1620s, shares both attributes. I wondered in Chapter 1 whether Kepler's camera might not have been of essentially the same design, with overhead lens, as that shown by 's Gravesande. This is speculative, but if it *was*, then Kepler's camera would predate 's Gravesande's by nearly a century; and if it *was not*, but had a horizontal lens tube in the wall and a vertical drawing surface, then this configuration plainly did not create impossible difficulties for Kepler in making his topographical views of the Austrian landscape.

There remains Delsaute's third point, that no 17th-century author 'seriously considers' the need or desirability of artists to use the camera, and that none offers any practical instruction. I have already discussed how Delsaute and others have interpreted the passage from van Hoogstraten's *Inleyding* quoted in Chapter 2, and have suggested that this relies on a somewhat selective reading. As for such other advice as was intended for artists, Delsaute is hardly fair to the sources. He says that della Porta's extended account of how to draw and paint using the camera (see p. 10) is 'not directed towards painters' (but if not to them, then to whom?). And he overlooks completely those Italian and French authors of perspective treatises from the late 16th and early 17th centuries, such as Barbaro and Niceron, who provide explicit instructions about the use of the camera for architects and painters.

In his *Pratica della Perspettiva*, for example, Barbaro sets out quite clearly a series of practical steps.[5] First one must make a hole in a window shutter. Then one must obtain a convex lens—not a concave lens—and set it in the hole. All other sources of light must be blacked out. An image can then be received on a sheet of card. The sharpness of the image can be adjusted by moving the card away from the lens or towards it. 'Here you will see shapes on the paper as they are in reality, with all their shading, colours, shadows, ripples on water, birds flying, and everything else that is visible.' Depth of field can be increased by restricting the aperture of the lens. When all is ready, one can draw outlines, and apply tones and colours, following Barbaro's instructions as in the passage quoted in Chapter 1.

Niceron's *La Perspective Curieuse* (*Ingenious Perspective*), published in Paris in 1652, advertises itself on the frontispiece as 'A very useful work for painters, architects, sculptors, engravers and all others who are concerned with drawing.'[6] Three pages are devoted to 'An optical experiment which teaches perfect perspective'. Here Niceron explains how to set up a room-type camera, what kind of lens and screen to use, and how to adjust the focus. 'So that if a painter imitates all the shapes that he sees, and if he applies to them the colours that appear so vividly, he will have a perspective as perfect as one could reasonably desire.' With a big spherical lens it is possible, he says, to obtain larger and higher-quality images than with a spectacle-sized lens. The inverted image may be righted by means of combinations of lenses, or with mirrors. Niceron also mentions portable cameras, both small—like portfolios [*portefeuilles*] or lanterns—and large—like military tents [*pavillons de guerre*].

There is, furthermore, the philosophical and mathematical literature, which cannot at this period be so clearly marked off from texts published for artists. Several perspective manuals were produced by mathematicians. Kircher's *Ars Magna* was widely circulated. Delsaute quotes Kepler as saying that he used the camera 'as a mathematician, not a painter,'[7] but maybe the astronomer is just being modest here about his own artistic talents, rather than expressing any opinion as to the limits of the instrument's applications.

It is true that there are very few surviving 17th-century drawings (and no paintings) for which there is documentary proof that they were made using the camera—one notable exception being Christopher Scheiner's drawings of sunspots in *Rosa Ursina*. Nevertheless, the historical account presented in Chapter 1 refutes all three of Delsaute's arguments against Vermeer's use of the camera. Cameras of several types *could* be used for drawing, and were so, from early in the 17th if not late in the 16th century. *No* crucial innovation was delayed until the 18th century. (If any feature or element was absolutely essential, it was the glass lens, introduced in the mid-16th century.) And there *was* technical advice directed specifically to artists, published from the 1580s onwards, about how camera images could be traced, and how correct perspective drawings could be made with the camera's help.

Allan Mills, an astronomer and expert on early optics, has also questioned whether the camera obscura could have been usable by artists, including Vermeer, in the 17th century.[8] He argues that the quality of images produced by single lenses would have been so poor at this period, as to render them valueless for painters. Mills assumes that Vermeer's lens would have been no more than 4.5 cm in diameter, the size of a spectacle lens. He made a series of experiments with a modern lens of this size (of better quality glass than that of 17th-century lenses, but otherwise comparable), which he used to project images of a rectangular grid of lines drawn on a board. The results showed up some serious deficiencies. There were two main problems, both of them around the edges of the image. Here the lines became slightly curved, and the image became progressively less bright and poorer in focus—the effect known as 'vignetting'

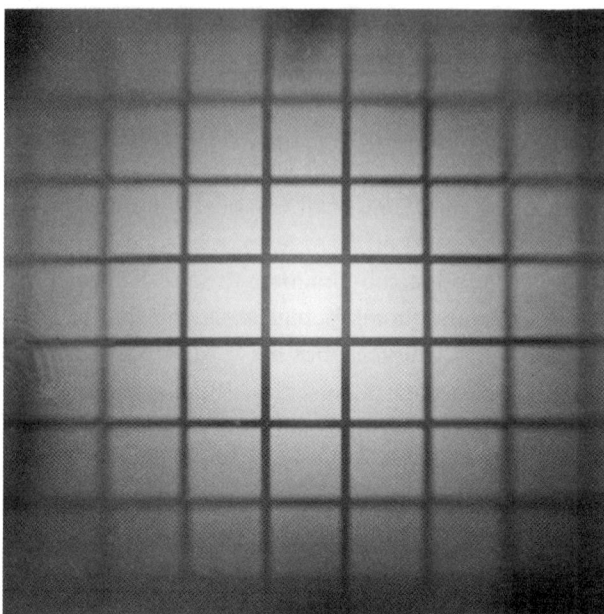

Figure 68
Photograph by Allan Mills of a rectangular grid of lines on a white board, taken using a simple convex lens. This illustrates around the periphery the effects of 'vignetting'—loss of brightness and focus—and curvature of straight lines, typical of camera obscura images generally.

(Figure 68). Mills also argues that such lenses would have had too limited a depth of field to allow the whole image of a room like Vermeer's to be in focus at once. Mills concludes that these difficulties would have been so great as to have made direct tracing of the images to achieve accurate perspective outlines a practical impossibility. Proponents of the contrary view could well have been misled by the images seen in modern cameras. Vermeer could only have obtained his perspectives by geometrical construction.

These points must obviously be taken seriously, but note their implications. The various failings of camera obscura lenses, which Mills demonstrates, were not overcome to any significant extent until the invention of photography. Only then was there the demand for all the optical improvements and sophisticated lens systems that occupied the camera industry well into the 20th century. Most of the camera obscuras made and used through the 18th and early 19th centuries had simple, single lenses not greatly different from those in 17th-century cameras.[9] Mills's position is thus open to the same kind of counter argument as Delsaute's: since cameras did not change very greatly in design over this long period, if they were usable later, then why not earlier? Set on one side for the moment any question about Vermeer: the logical conclusion of Mills's argument is that the camera obscura would have been of little use for tracing, or for observing fine detail in images, to all of the thousands of other users of the camera—painters, draughtsmen, commercial artists, and engravers—up to the 1830s. Were all these people unable to trace or pick out detail? Clearly not.

All this is not to say, by any means, that the images in simple camera obscuras are faultless. Problems of vignetting and loss of focus were clearly in evidence at the edges of the image produced by the cubicle-type camera in the BBC reconstruction—they occur in all single lens cameras. But to believe that this renders them unusable is to underestimate the skills that most artists would surely have for interpreting, improvising, and compensating for any deficiencies. It is the mistaken idea that working from a camera obscura is like taking a photograph which leads—I would suggest—to such a misconception.

I have mentioned earlier how it is perfectly possible for a user to readjust the camera obscura—working from inside, in a booth- or tent-type—to bring different parts of a scene into sharp focus at different times, where a photographer would want all parts focused simultaneously. The most serious of the image distortions produced by a simple uncorrected lens occur, as Mills shows, around the edges. But the effects of vignetting can to an extent be overcome in the camera obscura by refocusing so that the periphery is in focus (and the centre not)—although brightness is still lost towards the edges. This is achieved by moving the lens away from or towards the projection screen. A further possibility is for the lens to be *tilted* (while the screen remains fixed). This has no effect on the perspective, but can bring a band right across the entire image into sharp focus, following what is known to photographers as the Scheimpflug rule. If the lens is tilted about a horizontal axis this band is vertical; if tilted about a vertical axis, the band is horizontal.

What is more, in the image cast by a large lens, it is possible to select just a central section for copying—as I believe Vermeer did (see for example Figure 60). Any remaining slight curvature of what should be straight lines near the edges of a composition can be rectified by hand in the tracing or on the canvas. Camera obscura images are not perfect. But they *are* usable. It should be remembered too that the main reason for people first imagining that Vermeer might have used a camera was precisely that he appears to reproduce some of its optical deficiencies.

As mentioned earlier, some critics (including Mills) have based their doubts in part on the supposition that Vermeer's would have been a small box camera. Its screen would have been too small to trace images at the size of the paintings—with the exception of the two Washington panels—and its image too faint to give sufficient detail or be used indoors. The historical evidence on larger cameras and the full-size reconstruction built by the BBC argue strongly against this position. That demonstration depended crucially, nevertheless, on using a lens of something like 10 cm diameter—four or five times the area of those tested by Mills in his experiments. Were lenses of such a size readily available in Holland in the 17th century?

The answer is yes. Some were made as conventional hand-held magnifying glasses. One such can be seen on the table in the Verkolje mezzotint portrait of Leeuwenhoek (see Figure 19). It looks to be about 8 cm in diameter. Magnifying glasses were commonly used in the Dutch cloth trade—in which Vermeer's father had worked—for inspecting the weave of delicate fabrics. Otherwise large lenses were being made, increasingly as the century progressed for use in telescopes. Even as early as the 1580s an English writer, William Bourne, had described the manufacture of 'perspective glasses' up to 12″ (30 cm) across.[10] Constantijn Huygens, according to Monconys, possessed telescope lenses of 13.5 and 32.5 cm diameter.[11] The Royal Society has in its collection today three telescope lenses made by the younger Constantijn with diameters of 19.5, 21, and 23 cm, as big as dinner plates, all of which are dated 1686.[12]

Telescope lenses of these dimensions are specialized pieces of equipment, of course, and those made by the Huygens brothers, although not without their faults, were probably the finest of their time. (Also, the Royal Society lenses were produced after Vermeer's death.) It is nevertheless clear that the technology and skills *were* available in the middle of the century for grinding lenses of the more modest size that Vermeer would have needed. We know that lenses were being made commercially in Delft when Vermeer was working. Montias, for example, mentions one Evert Steenwijck, a spectacle-maker and lens-grinder in the city, who was producing telescope lenses as early as 1628.[13] Evert was the father of Pieter Steenwijck, an acquaintance of Vermeer's father. Further, the Huygens brothers were in correspondence with Johan de Wyk, another lens manufacturer in the city, in the years 1664 and 1665.[14]

One of the consequences of supposing that making pictures with a camera obscura is like taking a photograph is the idea that images copied from it must be like snapshots, catching whatever happens fleetingly to be in front of the lens. Thus Arthur Wheelock says that the composition of *The Little Street* suggests the use of the camera, because the buildings are cut by the frame at 'architecturally inexplicable places'.[15] Wheelock uses this as a point in favour of the camera hypothesis, but elsewhere he turns the argument the other way round—to contend that the extreme refinement and purification of Vermeer's compositions show that the painter used optical apparatus only as a source of first visual impressions.[16] The paintings, that is, are much too carefully considered to be 'snapshots'. But this is to miss the ways in which the camera could have served Vermeer in that very work of refinement and purification. The camera has the power to transform the three dimensions of any scene instantaneously into a two-dimensional arrangement of shapes on the viewing screen. It is an instrument with which the finest points of composition can be weighed.

It is certainly true, with reference to paintings like *The Little Street* or *View of Delft*, that a topographical artist cannot change the buildings which are in front of him. If he wants to manipulate their positions or shapes, he can do this only on his canvas. But the same does not apply to a spatial arrangement of figures and furniture like one of Vermeer's interiors, which can be deliberately set up, using the camera as an aid to composition. Vermeer could have provisionally chosen his viewpoint, positioned his camera, and roughly arranged the furniture and sitters. He could then have embarked on a prolonged process of making adjustments, altering the postures of his subjects, shifting the alignments of chairs and tables, arranging the exact fall of tapestries and carpets, maybe moving the camera itself—all the time returning to the camera image to study the resulting perspective occlusions, juxtapositions of shapes, the shapes of gaps or 'negative spaces' between objects, positions of shadows and brightly lit areas—until he was finally satisfied.

This is not to suggest that Vermeer never made further changes once he had started painting. There is plenty of evidence to the contrary—from *pentimenti* (changes of outline, visible through the final layers of paint) and from the revelations by X-ray and infra-red analysis of other changes of mind at different stages in the painting process—but none of this is incompatible with the use of the camera. Vermeer might have made late adjustments to the positions of actual objects in his studio, or he might have made drawings of objects in their initial positions, and then simply shifted their images on the canvas.

Swillens asked whether 'Vermeer had before him a given reality, that is to say, the complex of things and phenomena as they reveal themselves in his works', or whether instead he had 'separately drawn the things and phenomena and afterwards united them into a particular composition . . . born of his fancy and imagination.'[17] There is a logical difficulty in the way of answering this question, which was touched on in Chapter 5. In Vermeer's interiors there are identifiable maps, paintings, and pieces of furniture, as we know, which can be compared with the real objects where these still exist. But otherwise there is only the evidence of the paintings themselves as to what might actually have been in front of Vermeer's eyes. We cannot compare image with reality, because we have only the image. Here Swillens's question is, strictly interpreted, unanswerable.

Swillens recognized the objection and reformulated the question to ask rather are '. . . the light effects, perspective, colour and so on . . . always in accordance with the laws of science'? That is, are the paintings *internally consistent* in these respects? Any inconsistency—any departure from 'the laws of science'—would argue for a method which was not literally and painstakingly naturalistic. Swillens found few such inconsistencies. He concluded, after a detailed and lengthy analysis, that '*Everything [Vermeer] painted he saw immediately in front of him.* This is proved by: 1° the highly accurate representation of each kind of material, 2° the precise reflection of shadows and half-shadows, 3° the reproduction of details such as nails and their shadows and similar things in the wall, parts of furniture etc., 4° the exact, sharp drawing, 5° the careful way in which he grouped figures and objects and arranged their lighting.'[18] It was Vermeer's perspective accuracy and realistic detail, that is to say, which proved convincing for Swillens.

Fink, although differing from Swillens in subscribing to the camera theory, came somewhere near to Swillens's position on Vermeer's faithfulness to appearances. He pointed to Vermeer's consistency in representing the same architectural features and pieces of furniture from one painting to another. 'It does not seem probable that this kind of precise rendering . . . could have been possible without the use of the camera obscura.'[19] Even the fact that Vermeer copies artefacts of the camera ('actual appearances', that is to say, only on the screen, not in the real world) is a kind of paradoxical proof to Fink of Vermeer's tendency to reproduce always what he sees. Fink acknowledged the existence of *pentimenti*. Vermeer 'was not a slave to the camera obscura', and was prepared to ignore its image on occasion for compositional reasons, but the whole tenor of Fink's argument is that the basic outlines and much of the detail of the majority of Vermeer's later works are copied point by point from camera images. 'If the painter was convinced that the image on the ground glass of his viewing screen was the most satisfactory two-dimensional representation he had seen, it was unlikely that he would eliminate any facet of that image in making a painting.'[20]

These kinds of views have been dismissed by some critics as 'naïve realism', as 'obscura literalism'.[21] It is 'One of the many misconceptions about Vermeer's painting style that has affected theories of his use of the camera obscura', says Wheelock, '. . . that Vermeer was a realist in the strictest sense, that his paintings faithfully record models, rooms, and furnishings he saw before him.'[22] I do not agree completely with Swillens and Fink in their insistence on Vermeer's exact truthfulness at all times to natural appearances, but it is an obvious implication of my argument thus far that I share much in common with them. The very fact that it is possible to reconstruct the architecture of Vermeer's room and find so many close correspondences between the dimensions of windows, furniture, and decorations as they appear in different pictures provides ample confirmation for the claims which Fink and Swillens make for the painter's geometrical consistency. The fact that Vermeer depicts the maps, the paintings, and the identifiable furniture at their actual sizes—in many, if not in all cases—is further support for the idea that 'he paints what he sees'. (I am speaking here of truth to shape and outline—not necessarily the minute imitation of texture or surface pattern.)

There are other more subtle clues. Take a passage like the detail from *The Glass of Wine* shown in Figure 69. Here we can see past the nearer legs and rails of the chair, to the legs and rails on its far side; beyond them to the leg and rail of the table; and through the gaps between all of these, to the tiled floor surface beyond. This is an absolutely virtuoso piece of perspective rendering, geometrically correct in every detail. Anyone who has set up perspectives on the drawing board will recognize what technical difficulties are presented by these kinds of glimpses of distant detail through apertures in nearer objects. And yet this occurs in a part of the painting that is not enormously important to the composition as a whole. Even as accomplished a perspectivist as de Hooch would probably have improvised here, or shrouded the area in shadow. With an enormous amount of patience, the passage certainly *could* have been constructed geometrically. But to me, the fact that Vermeer gets it so right speaks of

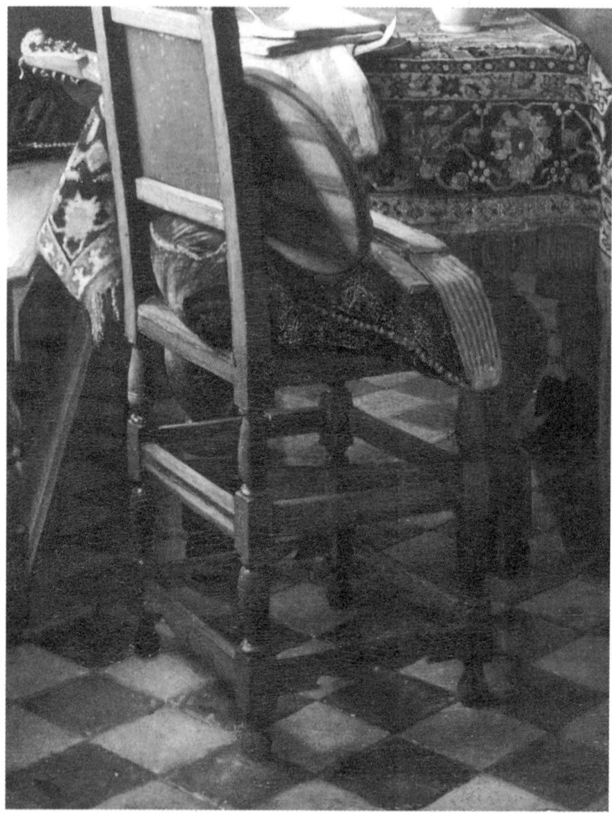

Figure 69
Detail from Vermeer, *The Glass of Wine*, c.1658–60, Staatliche Museen Preussischer Kulturbesitz, Gemäldergalerie, Berlin-Dahlem. Oil on canvas, 65 × 77 cm.

extremely careful observation of the real thing—whether directly or, as I believe, in a camera image. There are some comparable passages in other paintings: beneath the seat of the gentleman lutanist in *The Concert* and beyond the stockinged right leg of the artist in *Allegory of Painting*.

Another example: the image reflected in the hanging ball in *Allegory of the Faith* is very loosely painted, merely a series of coloured blobs (compare Figure 53a),[23] but the photographic reconstruction has shown just how accurately these correspond in their geometrical configuration to the layout of the room, transformed in shape by the curvature of the mirrored surface. This would be an almost impossible feat of mathematically constructed perspective. The reflection *must* have been observed in a real ball, suspended in a real room with the specific pattern of windows in question.

I would most emphatically not want to argue that Vermeer followed literal appearances at every point. I have already suggested (following Swillens) how he might have taken some artistic licence in the varying patterns of floor tiles, and in the painted lids of the virginals in the two pictures in the National Gallery in London. Dirck van Baburen's *The Procuress* does not appear at its correct size or shape in *Lady Seated at the Virginals*. Daniel Arasse has pointed to changes made by Vermeer in his copy of Jacob Jordaens's *Christ on the Cross* in *Allegory of the Faith*, where he has eliminated two figures, Mary Magdalene and a man on a ladder behind the cross.[24] My calculations serve to indicate that the open casements in the near foregrounds of *The Glass of Wine* and *The Girl with a Wineglass* are not consistent in size with other views of the room, and suggest that Vermeer has shrunk them in order to prevent them appearing excessively dominant. And so on.

On the other hand, there have been claims made, by Arthur Wheelock in particular, in criticism of 'naïve realism', about how Vermeer departs from correct perspective or from consistent rendering of light and shadow, which are highly debatable. Wheelock says, for example, that Vermeer has, for compositional reasons, adjusted the lid of the virginals in *The Music Lesson* to be slightly wider to the right of the girl than at the left.[25] I have measured the image and do not find this effect.

On light and shadow, Wheelock argues that Vermeer has artificially contrived the pattern of illumination in *The Music Lesson*, so that it does not obey the laws of optics. 'That direct sunlight enters the room is clear from the bright rear wall and pronounced diagonal shadow falling from the windowsill. Logically, however, sunlight that would create such a shadow would form similar ones at the juncture of the windows and the ceiling and behind the horizontal frame at the top of the lower window. Such shadows do not exist in this painting.'[26] According to Wheelock, Vermeer has selectively eliminated shadows, or has adjusted their angles to align them with significant points in the composition. He has made the shadows of the virginals' legs 'closer to each other than would appear in reality', and has 'eliminated the shadow of the instrument's body against the rear wall.'

The photographic reconstruction of the painting, however, produces patterns of light and shadow that are very close to Vermeer's original throughout (Plates 5 and 6). Recall that the model was lit by large diffusers to simulate north light—*not* direct sunlight. It is this fact, that the illumination comes from large areas and not a single point, which causes the shadows to strike at widely differing angles. There is one inconsistency which Vermeer *does* introduce here—in the angles of the shadows to the right of the mirror, relative to the angle at which the mirror actually hangs. Otherwise he shows a pattern of illumination which is perfectly logical.

View of Delft is a painting whose qualities have led many to suspect the use of the camera (Plate 3). Kenneth Clark described it as 'the nearest . . . painting has ever come to a coloured photograph'.[27] I have nevertheless given no attention to its perspective geometry, or to how true it might have been to actual appearances. This is because demolition and new building since the 17th century have made comparisons with the present scene almost useless. Figure 70 shows a recent photograph taken from the painting's viewpoint. Both the Schiedam and Rotterdam Gates, the most prominent buildings in *View of Delft*, were demolished in the 1830s. Many of the houses have also gone. A road now runs where the fortifications stood. Even the spire of the Nieuwe Kerk, caught in a shaft of sunlight in the painting, was rebuilt in the late 19th century.

Wheelock nevertheless insists that Vermeer was 'not concerned primarily with topographical accuracy'[28]. The large building at the left of the painting, he says, is 'almost certainly imaginary', the forms of the Rotterdam Gate have been flattened and distorted in perspective, the tower of the Nieuwe Kerk has been reduced in size, and so on. Because of all the changes it is obviously difficult to test the validity of these claims, but some of Wheelock's main supports are broken reeds to lean on. For example, Wheelock compares Vermeer's picture with an early 18th-century pen drawing of the same vista by Abraham Rademaker, and with diagrammatic buildings shown on the site in the 1675–8 pictorial map of Delft, and shows that Vermeer differs from both of these.[29] But why should we trust the testimony of these third-rank artists over that of Vermeer? Certainly when other topographical sketches by Rademaker are compared with scenes in Delft which are little altered over three hundred years, he turns out to be a quite unreliable witness.

All these are relatively minor details, and would hardly be worth quibbling about, if they were not adduced in support of a general argument about Vermeer's 'willingness to adapt reality' in order to achieve compositional balance. Certainly, Vermeer was prepared to depart from appearances—but by no means so freely or frequently as some critics have maintained.

I have set aside up to this point the possibility that Vermeer might have

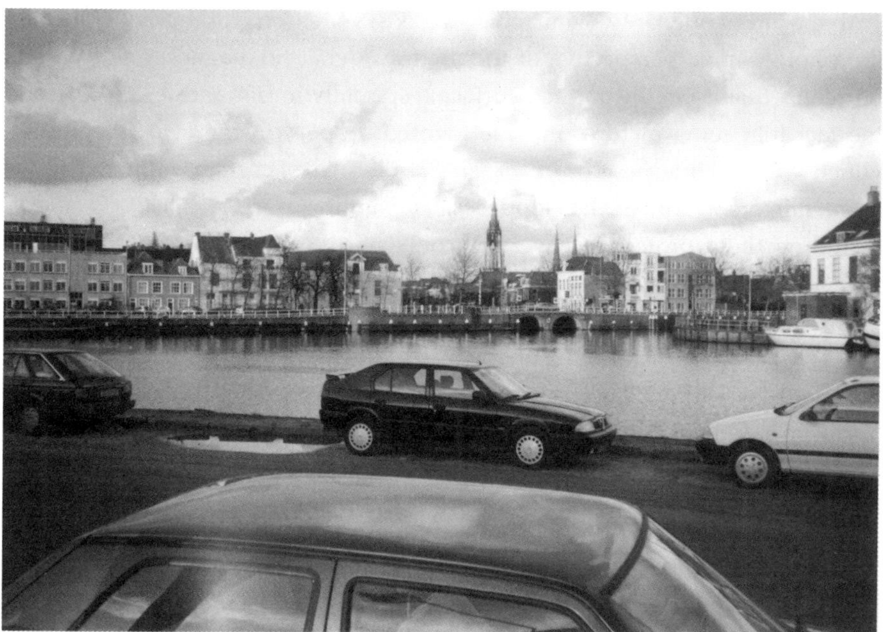

Figure 70 Photograph from the vantage point of *View of Delft*, as it appears today. Compare Vermeer's painting in Plate 3.

obtained his perspectives by other means than the camera. It is time to look at these other techniques. There is only a limited number of practicable alternatives. Vermeer might have worked geometrically using the standard methods set out in handbooks of perspective technique. He might have used mechanical aids such as Alberti's veil, the various gadgets illustrated by Dürer, or other machines following similar principles.[30] The 'veil' is a semi-transparent cloth screen worked with a grid of threads, set up between the artist and his subject, which allows him to plot what he sees onto gridded paper or a gridded canvas. Dürer illustrates a version consisting of a grid of wires in a wooden frame. Dürer's other devices achieve the same objective by letting the artist use stretched strings and sighting tubes to line up points in the scene with their images in the drawing or painting. Finally, it has been suggested by Wilenski and others that Vermeer might have traced over images reflected in mirrors.[31]

To take the mirror hypothesis first: Vermeer, according to Wilenski, 'frequently worked out his compositions and perhaps actually painted his pictures with the aid of one, or two, mirrors.'[32] Wilenski's reasons for this belief include the photographic perspective of the pictures; what he diagnoses as a certain 'vitreous' quality in the painted surface (especially in *Girl with a Red Hat*); and the fact that some pictures include painted drapery in the near foreground, which Wilenski interprets as *trompe-l'œil* renderings of curtains hung in front of large mirrors. Vermeer, he supposes, is painting mirror reflections and is including these protective coverings, only partly drawn aside. (Other more plausible explanations of these drapes were given in Chapter 2.)

For example, what most people would read as a doorway in *The Love Letter*, Wilenski sees as a mirror hung on the wall of the room in which the woman and her maid are sitting. The tapestry tied back at top right is the mirror's cover. Vermeer is hidden behind the right-hand edge of the mirror (as we view it). Wilenski even suggests that Vermeer has substituted the leaning broom, at the last minute, for the reflection of a protruding leg of his easel.

It is not as difficult to trace over the image in a mirror as one might immediately think.[33] One must close one eye, and fix the position of the open eye with a peephole or something like a gunsight. It might seem that one's head would get in the way. It does, but this is not so serious a problem. The image of one's head in a mirror has only half the dimensions of the real head. (Another difficulty is of course that the resulting image is reversed left-to-right. This is perhaps why Wilenski talks about Vermeer sometimes using a second mirror, to correct the image again.[34]) Whether Wilenski thinks that Vermeer worked like this is not entirely clear. He implies that Vermeer *studied* mirror images, not necessarily that he *measured* them. Certainly he does not seem to imagine that Vermeer traced mirror images at the actual sizes of the paintings; since large mirrors would at that time, he says, have been rare and expensive. (Nevertheless, the mirror in *The Music Lesson* measures about 60 × 70 cm according to my reconstruction, very nearly the size of the painting itself.)

Whether a technique of painting from mirror reflections is at all plausible historically in the case of Vermeer is not, however, the issue here. The question is could it account for the geometrical results of Chapter 6? Suppose for the sake of argument that Vermeer had had a mirror sufficiently large to reflect, at the actual sizes of the pictures, the complete extent of the interior visible in the six paintings in question. Figure 71 shows the positions, in the plan of the room, where Vermeer would have had to place this mirror. They are all in different planes. What is more, for the image to be full-sized, the mirror must in some

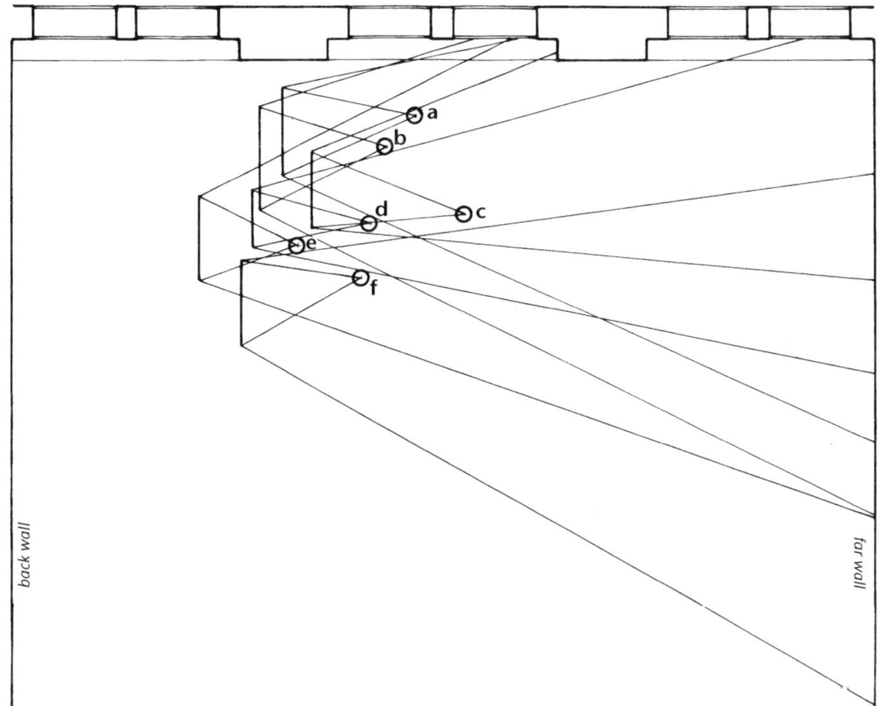

Figure 71 Positions in the room at which Vermeer would have had to place a mirror, to trace images of six paintings at their actual sizes. The circles show the viewpoints of the pictures: (a) *The Girl with a Wineglass*, (b) *The Glass of Wine*, (c) *Lady Writing a Letter, with her Maid*, (d) *Lady Standing at the Virginals*, (e) *The Music Lesson*, and (f) *The Concert*.

cases be a metre or more away from the viewpoint—too far to reach and trace with an outstretched arm. There is no means, that I can see, whereby a mirror technique could explain the special geometrical properties of the perspectives and their relation to the back wall. Similar arguments apply to Alberti's veil and Dürer's equivalent devices. A veil or grid would have to be placed, for each picture, in the same position as a mirror, to produce a full-sized image.

What of conventional mathematical methods? Swillens, who made a special study of Vermeer's perspectives, was not dogmatic about the artist's precise technique. 'The elements of perspective are all extremely accurate. . . . It cannot be decided with certainty whether he constructed these mathematically, and if so, what methods he used, or whether they are purely the results of exact observation.'[35] Other students, says Swillens, may have speculated that Vermeer resorted to artificial aids such as 'the painter's mirror, the net-frame, or similar implements', but such theories remain unproven.[36] On balance Swillens was inclined nevertheless, it seems, to think that Vermeer used geometrical construction methods. He mentions a number of early 17th-century perspective manuals to which Vermeer might have had recourse, including works by Hans Vredeman de Vries, Hendrik Hondius, and Samuel Marolois.[37]

This line of argument has been pursued more recently, and more vigorously, by Jørgen Wadum, who is convinced that Vermeer followed standard textbook perspective procedures and did not rely on the camera.[38] For example, Wadum suggests that some of the 'five books in folio' and 25 'other books', mentioned in the inventory of Vermeer's estate, might have been manuals such as those cited by Swillens. (Although perfectly possible, this is entirely hypothetical, as the document gives no indication of the books' titles or subjects.)

Wadum illustrates perspective analyses of several paintings, showing the positions of the central vanishing points, horizon lines, and distance points, similar to my own diagrams of *The Music Lesson*. He shows that the distance points are closer to the vanishing point in some early interiors, and move further out to left and right in certain later works. This implies that the early pictures are 'wide angle' views, and that Vermeer tended to take slightly narrower views as his career proceeded. Wadum speaks of the two-dimensional geometry of each picture as though Vermeer is deliberately controlling it by manipulating the positions of these respective points. In effect he is assuming what he wants to demonstrate: that Vermeer worked mathematically, setting out points and lines on the canvas, or in a preparatory drawing, to establish the basic outlines of the scene.

It is important to appreciate, however, that the existence of vanishing points and distance points in a perspective image is not in itself evidence that the artist has worked mathematically. A photograph of a comparable interior, viewed frontally, will have a vanishing point, a horizon line, and (if the floor has diagonal tiles) two distance points, in exactly the same way. So will an image traced in a camera obscura, or an image transcribed using a grid in a frame. These are intrinsic features of the optics and perspective geometry of the situation. The various points and lines are moved in the picture by the simple act of taking a different viewpoint, higher or lower, nearer or further away. Wadum proposes, for example, that 'Vermeer created his spatial illusions using the *tiers point* [vanishing point plus two distance points] method, first described by the French mathematician Pélerin.'[39] He bases this claim on the fact that, in Vermeer's pictures, 'the central vanishing point is placed midway between the distance points.' But this geometrical property is true of *any* 'frontal' perspective image, no matter how it be produced, whether mathematically, mechanically, or optically.

The one piece of firm evidence relating to Vermeer's techniques which Wadum brings to bear from the paintings themselves—and they are the only available source of information—is the fact that many of the canvases have tiny holes at the central vanishing points. K G Hultén was the first person to detect such a hole, in 1949, in *Allegory of Painting*.[40] Recent research has revealed more examples.[41] In *Allegory of the Faith* the hole is visible to the naked eye. In other cases a little of the lead white ground beneath the surface layers of paint has been displaced, and the result is to produce a black dot under X-ray photography.

There is really only one plausible interpretation of this phenomenon: that Vermeer put pins in the canvases at these points. Wadum suggests, very reasonably, that Vermeer might have attached a thread covered with chalk dust to the pin, stretched it tight, and snapped it to the canvas to make marks corresponding to the converging perspective lines.[42] This process might have left no traces (and none remain) in or under the finished paint surface, since the excess chalk could simply have been brushed away. Another possibility is that Vermeer might have set a straight edge against the pin and ruled the lines. Again Wadum interprets all this as positive confirmation of a mathematical technique of construction: but it is not.

Suppose that Vermeer was transferring an accurate preparatory drawing to the canvas by pouncing or some similar method (as Wadum imagines he might have done). It would have been useful to have a pin to help rule the converging lines over the top of the transferred impression, to get them absolutely straight and make them meet at a single point—no matter how that drawing had been produced. I have proposed that Vermeer might have made tracings on paper with the camera, and then transferred these to canvas. A pin at the central vanishing point would have been as useful here as it would for perspective drawings set out mathematically. I have found it helpful myself, when making perspective analyses on tracing paper over reproductions of paintings, to stick pins in at the vanishing points.

The idea that Vermeer might have set up his perspectives mathematically is in itself a reasonable one. The problem is that there is no positive evidence for this view, other than the geometrical correctness of Vermeer's rendering: no measured drawings, no written accounts by Vermeer or others, no certain knowledge that Vermeer studied or even owned perspective texts, not even—*pace* Wadum—the fact of the pinholes. Had it been possible to discover traces of perspective construction lines drawn on or scored into the preparatory grounds of the canvases, or any other tell-tale signs of mathematical working, then this would have been a different matter.[43] But none have been found: recall from Chapter 6 that no black or dark-painted outlines of any kind have been detected in the underlying layers of Vermeer's paintings.

It is important, what is more, to re-emphasize the *extreme* sophistication and precision of the underlying perspective geometry of Vermeer's pictures. I have already referred to the technical difficulties posed to a mathematical method of working by certain passages—the legs of chairs, the mirror reflections. Wadum imagines Vermeer, like some contemporaries, using the simplest of methods—'placing a pin attached to a string through the canvas'[44]—as though the only problem involved was that of drawing the images of receding orthogonal or diagonal lines. The real difficulty on the contrary is that of precise *measurement*.

The reconstructions detailed in Chapter 5 have demonstrated that Vermeer, had he worked mathematically, would have needed to adopt some much more painstaking and elaborate procedures. His images of real items of furniture and real maps show them at their precise actual sizes, allowing for diminution with distance. The dimensions of the room and of recurring items of furniture, as we have seen, are represented consistently from one painting to another. In order to achieve such results by following mathematical methods, Vermeer would have needed to prepare accurately measured working drawings both of the room itself and of all the items depicted in it. A geometrical technique is conceivable, but it would have been far from simple. Some of the more complex construction drawings in Vredeman de Vries's *Perspective* or Hondius's *Institutio Artis Perspectivae* (*Principles of the Art of Perspective*) give an idea of what would have been involved.[45] (On the other hand, all this perspective precision and dimensional consistency would have been given directly by the careful transcription of images in a camera obscura.)

If the central vanishing points of Vermeer's pictures were to coincide precisely with meaningful points in the compositions—in the eyes or hands of the figures, say, or on especially significant objects such as love letters, sheet music, or tempting glasses of wine—then this again might hint at a mathematical method of construction. The perspectives could have been set up deliberately to achieve such results. With a camera, by contrast, the position of the vanishing point tends to fall more fortuitously (although a camera *can* be lined up with care to centre on some chosen point, if required). Some writers have indeed argued that Vermeer's vanishing points are deliberately and artfully positioned, as for example in *Lady Writing a Letter, with her Maid*, where the point is at the centre of the lady's face. In relation to other paintings, however, the argument is unconvincing. In *The Music Lesson* the point falls on the woman's left elbow. In *Allegory of Painting* it is on the knob at the left-hand end of the rod on which the map of the Netherlands is stretched. In *Allegory of the Faith* it lies in an area of blank wall between the tapestry and the painted 'Crucifixion'. The vanishing point of *The Love Letter* lies some way to the right of the doorway through which the subject of the picture is seen. None of these seem especially significant in relation to the painting's compositional or dramatic structure.[46]

Meanwhile the key question remains: could a mathematical perspective method account for the curious geometrical phenomenon reported in Chapter 6? This is the decisive issue. If Vermeer had worked mathematically, why would the 'projected images' on the back wall of the room work out at the exact sizes of the paintings themselves? I can think of no convincing reason. One possible argument might be that Vermeer had a single standardized perspective image of the walls and floor of the room, with a fixed viewpoint, which he used as the common underlying basis for a whole series of pictures, selecting now this area of the 'master image', now that. The geometrical consequence would indeed be for all the 'projected images' to coincide in one plane. (For they would just be different parts of one and the same image.) However, the theoretical viewpoints of Vermeer's interiors are *not* all identical (compare Figure 49). On the contrary, they differ significantly in their positions relative to the far and side walls of the room, and the detailed perspective structures of the various paintings vary as a consequence; so this idea is unsustainable. I can think of no plausible explanation as to why any mathematical perspective method should produce the results of Chapter 6, which are so straightforwardly and simply accounted for by a camera technique.

9 The influence of the camera on Vermeer's painting style

If Vermeer was indeed a user of the camera as I have argued, what consequences did this have for his method of applying paint, his visual language, even his choice of subject matter? Thus far I have concentrated on technicalities of perspective geometry and on the objective realities (so far as they can be determined) of Vermeer's studio and its furnishings—an analysis which I hope has had some historical value in itself. But does this glimpse of what went on behind the scenes, inside Vermeer's rehearsal room, have any significance for what the audience experiences of the play itself? It might be argued that the paintings can be fully appreciated without any of this background knowledge.

I believe otherwise. I believe that Vermeer's special technique was intimately bound up with his special vision. This concluding chapter develops some tentative suggestions about the nature of this relationship, leaning heavily in places on the insights of other critics, in particular Lawrence Gowing. Let me re-emphasize at the outset that these are offered not as any kind of general critical appraisal, but just as some pointers towards the specific impacts on Vermeer's painting of a camera technique. Maybe to speak of 'impacts', indeed, involves too much of a technological determinism: perhaps one should ask, rather, what was it about the camera's potential that particularly appealed to Vermeer's character and sensibilities, and offered him a way to realize his artistic project?

I have already spoken in the last chapter about the camera obscura as a 'composition machine', and have imagined Vermeer moving back and forth between the room and his darkened cubicle, shifting furniture, rearranging drapes, asking his model to turn her head or fold an arm. Swillens, although agnostic as to Vermeer's methods of obtaining his perspective precision, conceived of something similar:

> Paintings are taken down from the walls and replaced by others or by maps. Chairs, covered with leather, or with ornamental material, are brought out and given a place in the whole. Windows are shut, others opened, curtains pulled forward, then again taken away and replaced by those of other colours. A musical instrument is given a place against a chair, as if a player had just laid it down ... The modelling woman takes the place appointed to her by the painter. He asks her to move a little to the left, then to come forward, and above all to stand still.[1]

Quentin Williams, referring to *Officer and Laughing Girl*, sees Vermeer and the woman working painstakingly together, each understanding their roles 'every bit as thoroughly as a director and actress in a film studio.'[2]

The camera collapses three-dimensional space onto the two-dimensional plane, it disables those faculties of depth perception which depend on binocular vision and parallax. (Objects at different distances no longer appear to move relative to each other as one moves one's head.) Solid objects become flat shapes, and their forms are comprehensible only by their outlines, their tonal modulation, and the shapes and positions of their shadows. The spaces that separate objects are also turned into flat shapes. All these shapes can then be studied in compositional terms, and can be manipulated by *moving the objects themselves*. If Vermeer's camera was such as to produce an upside-down image, then this would have created difficulties for him, admittedly, in judging the picture's proper appearance; it might curiously have had some virtue, at the same time, in forcing him to see the composition more abstractly in terms of relations between shapes. Professional photographers who use plate cameras find that they become accustomed to assessing compositional relationships in inverted images.

Some critics have attributed to his use of the camera Vermeer's tendency in a few paintings to cut objects off abruptly and arbitrarily at the pictures' edges. I have already cited Arthur Wheelock's diagnosis of this characteristic in *The Little Street*, which slices the two houses at both sides of the picture, right through the middles of their windows. It is hardly surprising that Vermeer never goes so far as, say, Degas in imitation of optical or photographic images, by truncating human figures at the sides of the canvas. But there is something of this effect, as Williams has pointed out, in the way that the table in *Young Girl with a Flute* and the chair back in *Girl with a Red Hat* are cut to leave so little showing at the bottom of each picture as to render these pieces of furniture almost entirely meaningless and unexplained.[3]

These are isolated examples, however. Much more generally it is, by contrast, the carefully measured and resolved proportioning and balancing of shapes, both positive and negative, which characterize Vermeer's compositions and contribute to their feeling of static repose and inevitability. Even where chairs or maps *are* cut by the frame, enough is shown to make plain what is depicted, and the very truncation seems to be deliberately designed to create rectangles of the desired shapes and sizes. Gowing speaks of Vermeer's canvases as 'mosaics' of shapes which 'bear equally on one another'. 'They are clasped together by their nature, holding each to every other in its natural embrace. We see a surface which has the absolute embedded flatness of inlay, of tarsia.'[4] It is this way in which Vermeer first carefully controls and adjusts, and then follows but simultaneously abstracts from the camera image, which provides an explanation for this otherwise perplexing and paradoxical quality: a perfect perspectival illusion of depth *coexisting* with an effect of surface flatness which can suggest mosaic or marquetry. As André Malraux remarks:

> Some have spoken of the 'recessions' in the *View of Delft* and the *Street in Delft*. Actually, when we examine the originals . . . we are struck by their lay-out in large planes perpendicular to the spectator.[5]

Vermeer's pictures have provoked comparisons with photographs, as we have seen, since the 1860s. But what does it mean, precisely, to describe his work as 'photographic'? There is the matter of perspective accuracy and the concomitant perspective distortions resulting from taking close-up or wide-angle views, which Pennell was the first to notice. Beyond this, a naïve conception of what is 'photographic' might put greatest emphasis on laboured, painstaking accuracy in the minute explanatory transcription of detail. If this was truly the distinguishing criterion, then it would be Gerrit Dou and the Leiden 'fine painters' who would better qualify as 'photographic', but unlike Vermeer, they are not usually so described.

The truth is that Vermeer manages to achieve 'photographic' results while painting in a way which is often locally imprecise, where focus is sometimes lost, where areas of colour may be simplified and flattened, texture obliterated. As René Huyghe says:

> it is misleading to praise Vermeer for his *painstaking art of detail and imperceptibility*. In fact no painting is broader than Vermeer's, with its prodigiously accurate but dissolved touch, almost anonymous, applied in spreading pearls, scorning petty precision, and which, viewed from close at hand . . . gives the impression of a slightly blurred image not quite 'in register' . . .[6]

Vermeer is true to tonal values, true to the light from objects seen indistinctly, true to the view through half-closed-eyes as it were, *not* always true to detail. As Gowing puts it, Vermeer's work has certain affinities with *trompe-l'œil* painting, but 'at the opposite pole to naturalistic tactility'. He creates an illusion 'not of closeness but of distance.'[7]

It is the camera obscura that constrains him to do this. The instrument inevitably introduces such a quality to an extent, in any case. It is impossible, with a single uncorrected lens, to achieve perfect focus throughout all parts of a large image. A ground-glass or translucent paper screen, if this is used, introduces its own further effects of softening and simplification. Suppose, however, that we stop thinking of these as defects and obstacles, and imagine instead that Vermeer might, up to a point, have actually valued these idiosyncrasies of the camera. In its defocused and 'condensed' form, the image can more readily be resolved into areas with uniform tonal and chromatic values, and the thresholds established at which these values change. It may then happen that some of these boundaries do *not* fall at the edges of objects in the scene. Vermeer starts to paint patches of light and colour, not fingers or bodices or violas with the forms and outlines by which they are mentally conceived.

This is the characteristic that, above all, led Gowing to suspect Vermeer of using the camera. This is the reason for the absence of conventional drawing. This is the explanation for the pattern of underpainting in simple areas of pure tone, pure lights and darks (Figure 16). In *Girl with a Pearl Earring* there is no line at all following the profile of the girl's nose on the left-hand side. The bridge of the nose is given precisely the same colour and tone as the cheek beyond. The lines of the right side of the nose and nostril are nearly lost in shadow. And yet we imagine and read an outline, and so a form, because of our prior knowledge and our reading of the shape of the shadow. The same is true for *Head of a Girl* in the Metropolitan Museum in New York.

Vermeer is happy—unlike many of his contemporaries—to depict familiar objects, or parts of the body, in positions or from angles where their forms are not clear from outline or shape. Gowing points to a number of examples in his

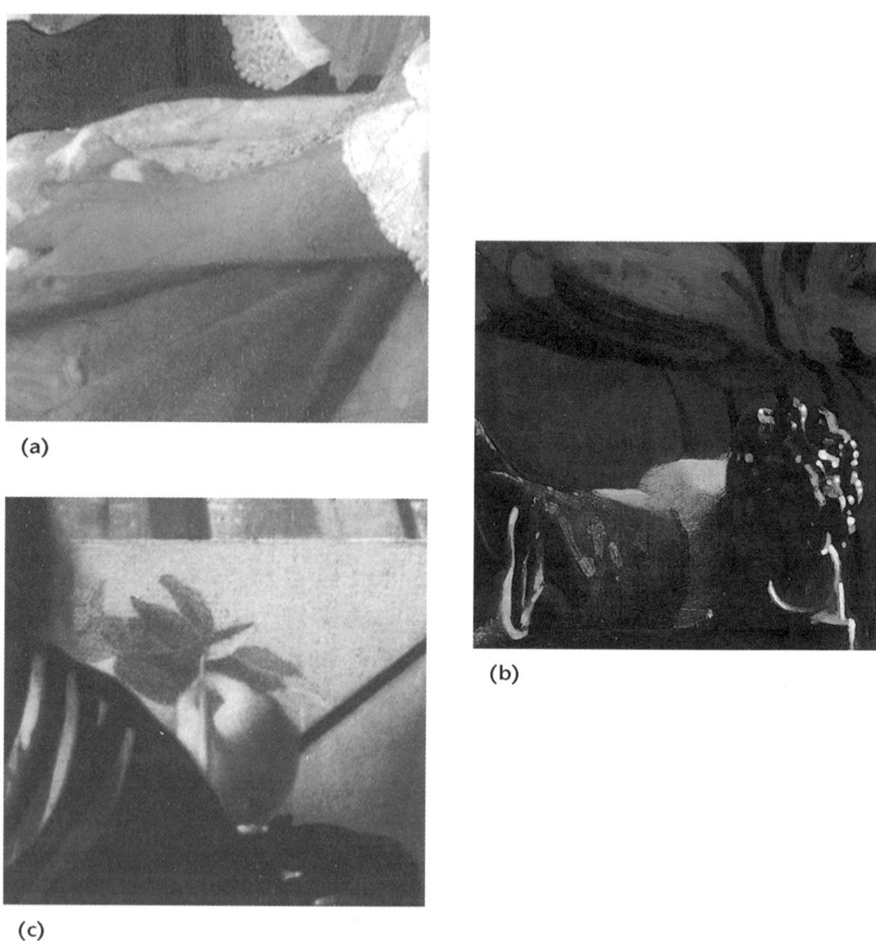

(a)

(b)

(c)

Figure 72 Vermeer's treatment of hands: (a) the left hand of the woman in *The Girl with a Wineglass*; (b) the right hand of the figure in *Girl with a Red Hat*; (c) the painter's right hand in *Allegory of Painting*.

treatment of *hands* (Figure 72). The girl in *The Girl with a Wineglass* has her left hand in her lap. The colour and tone of hand and wrist are almost uniform, unmodulated. There is little or no line. As Gowing says, 'Its foreshortening leaves all the receding surfaces unexplained. Yet it is entirely convincing. Its linear shape is insignificant, yet as a record it is entirely and effortlessly complete.'[8] Likewise, Williams draws attention to the right hand of the figure in

Girl with a Red Hat: Vermeer has made it 'square and stubby by the pressure of his faith in [its] already transmitted reality—by his simply copying it out.'[9]

The painter's right hand in *Allegory of Painting* provides another example. We see it as one rarely sees a hand: from behind, looking directly along the forearm. 'By the hazard of [Vermeer's] inflexible impartiality', Gowing says, 'the hand takes on a fortuitous bulbousness which has few parallels anywhere in art.'[10] And yet Vermeer seems willing to set down the areas of light and shade just as they occur. Because we know what we expect to see, we can interpret them. We do not—unless our attention is specially drawn—find them very remarkable.

> The description is always exactly adequate, always completely and effortlessly in terms of light. Vermeer seems almost not to care, or not even to know, what it is that he is painting. What do men call this wedge of light? A nose? A finger? What do we know of its shape? To Vermeer none of this matters, the conceptual world of names and knowledge is forgotten, nothing concerns him but what is visible, the tone, the wedge of light.[11]

When Kenneth Clark compared *View of Delft* to a colour photograph it was precisely these qualities that he had in mind: an 'uncannily true sense of tone', deployed with 'an almost inhuman detachment.'[12] The poet Paul Claudel gave a lecture on Dutch painting in The Hague in 1934 in which—before any of the empirical researches of Seymour, Fink, or Wheelock—he too attributed 'the photographic quality in Vermeer's paintings' to his use of the camera. It was not just Vermeer's colours that led him to this view, 'so exact and so cool' that the play among them seems to be 'not so much produced by the brush as realised by the intelligence'. 'What fascinates me', says Claudel, 'is a pure gaze, bare, sterilised, washed of all substance, of an ingenuousness in some ways mathematical or angelic, or let us say simply photographic, but what photography: in which this painter, a recluse hiding behind his lens, captures the exterior world.'[13]

I have argued, on the other hand, that to copy or work from an image in a camera obscura is distinctly *not* like taking a photograph. The process is not instantaneous but protracted. The focus in different parts of the image can be selectively adjusted. There are differences too, stemming from the fact that this image is *projected*, like a slide on a screen, or even backlit, and thus has a brilliance and animation—despite its dimness in absolute terms—which printed photographs lack. Many writers have described the special effects which are so produced: local colour becomes 'condensed' and concentrated, shadows are made relatively darker, contrast seems stronger. Martin Kemp has argued that Vermeer's 'exquisitely refined grading of simple and compound shadows in response to different kinds of direct and diffuse illumination is obviously the result of many hours of purposeful looking. It is easy to imagine that such an artist would have been attracted to the optical charms of the camera obscura.'[14]

It would be a mistake, therefore, to imagine that, because Vermeer's pictures might look to a degree like colour photographs, that his working process had many affinities with those of modern painters who have, precisely, worked from colour photographs. Quentin Williams makes this point in a perceptive essay on painting and photography. Copying from 'projected visible actuality' as in a camera obscura serves, he says, to produce a very different result.

> This is probably because it is, by contrast, a basically kinetic process: because moving figures shading or interrupting the image, or shifting clouds changing its light content, give the copyist a sensation very comparable with working directly from vibrant life itself.[15]

In this particular respect then it is perhaps wrong to categorize Vermeer as 'photographic'. He has managed to escape the inertia, the mechanical deadness which can come from too dutiful an imitation of the photographic print.

Can Vermeer's use of the camera be related to his choice of subject matter? To propose any direct causal relationship would be simplistic—as if the means determined the aesthetic purpose. However, if we are prepared to consider the matter the other way round, then we can perhaps speculate a little on what it was about the camera—both its potentialities and its constraints—that might have specially attracted the artist. Even so, it is necessary be cautious, since this involves attributing purposes and character traits to a man of whom we have no personal knowledge other than through the paintings.

At a simple mechanical level there is the fact that the perspectives of Vermeer's interiors are all 'frontal' or 'central' in their geometry. This would be a direct consequence of using a booth-type camera in which the back wall of the room acted as a projection screen. The picture plane would then be necessarily parallel with the far wall of the room. Not too much should be made of this, certainly, since frontal perspective characterizes so much of the rest of Dutch genre painting—by de Hooch, Steen, ter Borch, Metsu, and many others. Nevertheless, this is a time when the Delft architectural painters—Houckgeest, de Witte—are experimenting with oblique-angled perspectives, and the technique is even occasionally applied to domestic interiors, as in *The Gold Weigher* of Cornelis de Man. This Vermeer never attempts.

Vermeer's early subjects engage the viewer at close range in the psychological relations between small groups of figures—*Christ in the House of Martha and Mary*, *The Procuress*—with little or no architectural background. As he progresses through the latter half of the 1650s, however, and adopts a camera technique, he retreats from his figures and frames them with walls, curtains, and furniture that begin to be as much his subject as the human protagonists. The men and women themselves become progressively detached from one another, and it is ever more difficult to grasp what their emotional relationships might be—as in *The Music Lesson* or *The Concert*. This is not to say that emotions or psychological attachments are absent; rather that they are expressed more obliquely and ambiguously. Increasingly, Vermeer concentrates on one or two figures, usually women, absorbed in their own thoughts or domestic work.

From the earliest paintings, Vermeer reveals a sensibility at the opposite pole from the jovial chaos of Jan Steen and the thumping obviousness of his moral messages. André Malraux speaks of Vermeer 'tiring of the anecdote'.[16] There is even a gloom and anxiety about Vermeer's 'merry companies', *The Girl with a Wineglass* and *The Glass of Wine*. He seems happiest when his sitters are busy with their own activities, and he does not have to effect any social introductions. There are a few occasions in the early pictures where a figure looks directly out at us—the black-hatted gentleman at the left of *The Procuress* (some say a self-portrait, but he looks too blithe and gay), the silly woman in *The Girl with a Wineglass*—but later all the men and women look down, or turn away. It is only in the one-to-one intimacy of the little portraits that we and the painter meet his female subjects' gaze directly.

Can we relate these preoccupations and emotional tendencies to a camera technique? At the level of simple logistics, the camera user requires his human subjects to stay *perfectly still*, at least long enough for their faces and hands to be studied and drawn. (Perhaps at other times their clothes can be draped on lay figures.) The Vermeer household was continuously filled with young children—at least twelve born in fifteen years—but, with the possible exception of the portrait heads, none ever makes an appearance. We know from Montias that there were fights and arguments on occasion elsewhere in the house, but these never disturb the peace of Vermeer's studio.[17] There are none of Jan

Steen's inquisitive farm-dogs or van Mieris's copulating spaniels. Vermeer's servant girl might fall asleep, but there is no lurking cat, about to sample the pie or make off with the fish. (Richard Polak made some carefully composed photographic pastiches of Vermeers in the early years of the 20th century, several of which feature a cat,[18] but close inspection shows that it has the exact same posture in every case. The animal is *stuffed*: this, however ludicrous, seems curiously appropriate.)

As for Vermeer's adult models, their poses are almost all such as could be held for long periods without difficulty. Arms lean on tables, hands rest on stringed instruments or keyboards. Only the female figures in *Woman Holding a Balance* and *The Girl with a Wineglass* would have been under any real strain. As Albert Blankert puts it: 'In the real world figures move, but Vermeer's meticulous technique prevented him from creating an effective suggestion of movement.'[19] Vermeer is able to carry off the unusual feat of including human figures in what otherwise seem like *trompe-l'œil* pictures by placing them 'in circumstances where we expect no activity from them.' If the various models sat *separately*, at different times, then this might have served further to break the emotional contact between them. There is no speech in Vermeer's pictures. People communicate by letter, or through music. It has become a commonplace to speak of the stillness and detachment of Vermeer's interiors. One of the most poetic expressions of this feeling is from Charles de Tolnay:

> Vermeer's paintings can no doubt be defined as the most perfect still lifes of European art—still lifes in the original sense of the word, that is to say 'silent life', *Still Leben*, dream of a perfect reality, where the calmness surrounding things and beings almost becomes a substance, where the objects and the figures (treated as objects) give us to understand the secret relationship between them. Time here appears to be suspended, daily life takes on the guise of eternity.[20]

Connection to the external world of action, affairs, and learning is made by the men—occasional male visitors, unseen senders of letters, scholars who stare in imagination at the continents or the stars. Alone among the men it is the painter himself who is entirely at home, happy to be confined within the walls of his studio.

Even those critics who are doubtful about or hostile to the camera theory would agree about a pervasive sense of hesitancy and reticence in Vermeer's work. He is not indifferent to his subjects but he wants to contemplate them from a safe distance. He puts barriers between himself and his sitters—chairs, tables with heavy carpets over them, heavy tapestries. And then he retreats still further, into the darkness behind the screens of his cubicle, to become a voyeur through his lens. For Lawrence Gowing the technical and the personal become 'deeply tangled' here:

> His tonal method, his vocabulary of light, provided the solution. It was in the camera cabinet perhaps, behind the thick curtains, that he entered the world of ideal, undemanding relationships. There he could spend the hours watching the silent women move to and fro.[21]

If I too might be permitted a little personal fantasy, I imagine Vermeer taking a decision early in his career to confine himself to one room, and to create a whole world within it from the simplest of means: a few friends and members of his family, their best clothes, some of his family's collection of paintings and treasured pieces of furniture. It is clear from his small production that he worked slowly and was not commercially driven to repeat subjects endlessly, as de Hooch was. Perhaps his income from art dealing, or his mother-in-law's money, allowed him this luxury, until debts caught up with him. And then he disposed these simple elements, in a few pictures, to say what he wanted—however elusively—about the subjects for which he really cared: domestic routine, the love of men for women and fathers for daughters, the consolations of music, the worlds of science and scholarship, his own profession and its ambitions—all captured in his room within a room, his camera in a camera.

APPENDIX A:

Architectural features appearing in Vermeer's interiors

The paintings are listed in something approximating to what is believed to be their chronological order, but adjusted so that companion pieces or pictures with closely related subjects are placed together. Dates quoted here are as given in A K Wheelock (ed.), *Johannes Vermeer*, catalogue of an exhibition held at the National Gallery of Art, Washington and the Mauritshuis, The Hague; Yale University Press, New Haven Conn. (1995–6). The features include types of window, types of floor tile, and whether or not a ceiling is visible. The windows are classified into 'lozenge', 'hourglass', and 'squares and circles' types, as explained in Chapter 4. The different patterns of marble tiles are distinguished. The column labelled 'Swillens's rooms' refers to the five different rooms A, B, C, D, and E proposed in P T A Swillens, *Johannes Vermeer: Painter of Delft 1632–1675*, Spectrum, Utrecht and Brussels (1950), pp. 73–4. The final column, 'Possible to reconstruct?' identifies those paintings with a sufficient area of tiled floor to allow a three-dimensional reconstruction (see Chapter 5).

Appendix A: Architectural features appearing in Vermeer's interiors

	WINDOW TYPES				
	Lozenge	Squares/circles	Hourglass	[Unique]	Ceiling visib
1. *A Girl Asleep*, c.1657					
2. *Girl Reading a Letter at an Open Window*, c.1657					
3. *Officer and Laughing Girl*, c.1658					
4. *The Milkmaid*, c.1658–60					
5. *The Glass of Wine*, c.1658–60					
6. *The Girl with a Wineglass*, c.1659–60					
7. *Girl Interrupted at her Music*, c.1660–1					
8. *Young Woman with a Water Jug*, c.1664–5					
9. *Woman with a Pearl Necklace*, c.1664					
10. *Woman in Blue Reading a Letter*, c.1663–4					
11. *Woman Holding a Balance*, c.1664					
12. *Woman with a Lute*, c.1664					
13. *Lady Writing a Letter, with her Maid*, c.1670					
14. *The Music Lesson*, c.1662–4					
15. *The Concert*, c.1665–6					
16. *Allegory of Painting*, c.1666–7					
17. *Allegory of the Faith*, c.1671–4					
18. *The Love Letter*, c.1669–70					
19. *The Astronomer*, 1668					
20. *The Geographer*, c.1668–9					
21. *The Guitar Player*, c.1670					
22. *Lady Standing at the Virginals*, c.1672–3					
23. *Lady Seated at the Virginals*, c.1675		?		?	

TYPES OF FLOOR TILE

Ceramic	White squares on black ground	White crosses on black ground	Black crosses on white ground	Swillens's rooms	Possible to reconstruct?
				D	
				D	
				D	
				B	
				B	
				A	
				A	
				A	
			?	A	
			?		?
				A	
				A	
				A	
				A	
				A	
				A	
				E	
				E	
				C	
				C	

APPENDIX B:

Measurements of Vermeer's room and furniture

Table B1: Actual and reconstructed sizes of two maps and van Baburen's *The Procuress* (dimensions in cm)

	Actual size	*Size as estimated from reconstruction*
Map of Europe in *Woman with a Lute* (lower left part only, without text border)	109.8 × 117.0	114.6 × 124.8
Map of Seventeen Provinces in *Allegory of Painting* (central map section only)	111.0 × 153.0	105.0 × 149.4
Van Baburen's *The Procuress* in Vermeer's *The Concert*	100.8 × 107.4	91.8 × 108.6

Table B2: Reconstructed sizes of various items of furniture in 11 paintings
(dimensions in centimetres).
Note: chairs marked* are similar to the lions' head type in general design, but lack the lions' heads)

Painting	*Lions' head chairs*				*Tapestry chairs*			
			Seat in plan				*Seat in plan*	
	Height of seat	*Total height*	*width*	*depth*	*Height of seat*	*Total height*	*width*	*depth*
The Glass of Wine	48.3	101.6	35.6	35.6				
The Girl with a Wineglass					53.3?	97.8		
Woman with a Lute	51.6	105.2 110.7	40.4	40.4				
Lady Writing a Letter, with her Maid	50.8	99.1*	55.9	55.9				
The Music Lesson	53.3	112.8	44.2	44.2				
The Concert					53.1	97.3	49.3	41.4
Allegory of Painting	49.5	94.0*	39.4	39.4				
Allegory of the Faith	44.9	not visible	40.1	38.3				
The Love Letter								
Lady Standing at the Virginals								
Lady Seated at the Virginals								
Averages	49.7	107.6 96.5*	42.4	42.4	53.2	97.6		
Actual sizes of surviving examples	48.5	108.5	39.0	43.0	54.0	99.5	49.0	40.0

Notes on Tables B2 and B3

Table B2 shows the reconstructed sizes of the two types of chair (the lions' head and the tapestry-covered designs), the tables, the virginals, and the white Delftware jugs, wherever they appear in the 11 pictures listed. All the chairs of a given type are broadly consistent in their dimensions, with the principal exception of the measurements in plan of the seat in *Lady Writing a Letter, with her Maid*, which are unusually large. This chair is, however, rather hard to reconstruct precisely, since barely half of it is visible. In this painting and in the *Allegory of Painting* the chairs seem to be of the same general design as the lions' head type, but without the lions' heads (and with different upholstery). They are correspondingly reduced in total height.

As for the tables, those where the special design of leg is visible do not vary very greatly in either height or width (although the width can be measured only in two cases). The estimated lengths on the other hand are

Table B2 *continued:* Reconstructed sizes of various items of furniture in eleven paintings
(dimensions in centimetres)
Notes: tables marked § are the 'bulbous-kneed' design; average dimensions for the virginals exclude the instrument in *The Music Lesson*

Painting	*Tables*			*Virginals*			*White jugs*
	Height	*Width*	*Length*	*Height (lid shut)*	*Width*	*Length*	*Height*
The Glass of Wine	76.2§	76.2	114.8				22.9
The Girl with a Wineglass	73.7§	68.6	114.3				22.9
Woman with a Lute	83.1§	not visible	163.3				25.9
Lady Writing a Letter, with her Maid	73.7	48.3	not visible				
The Music Lesson	76.2	71.1	not visible	121.9	56.4	195.1	
The Concert	81.3§	not visible	194.3				
Allegory of Painting	76.2§	not visible	200.7				
Allegory of the Faith	67.1	not visible	not visible				
The Love Letter							
Lady Standing at the Virginals				98.0	52.3	154.9	
Lady Seated at the Virginals				106.7	47.0	207.8	
Averages	72.3 78.1§	59.7 72.4§	163.5	102.3	49.6	181.3	23.9
Actual sizes of surviving examples	80§ 88§	101§ 94§	175§ 260§		47.7 48.8 49.5	166.4 142.2 171.4	

much more variable, the longest being nearly twice the length of the shortest. These measurements are intrinsically unreliable, because of the table-tops being seen at *very* shallow angles. A tiny change in the angle at which a 'projector' is drawn in such circumstances (see Chapter 5) will result in a significant difference in the estimated dimension. Such very large discrepancies must, however, raise the question of whether this is indeed the same piece throughout. Another possible explanation may lie in the fact that tables of this design were often capable—as surviving examples show—of being altered in length by means of extendable leaves.

The two examples of these bulbous-kneed tables in the Prinsenhof Museum in Delft, referred to in the text, are similar in height (80 cm, 88 cm) to those which Vermeer paints. But they differ greatly from one another—as well as from Vermeer's table or tables—in both width and length (that is, length measured without the leaves pulled out). One measures 101 × 175 cm, the other 94 × 260 cm. Swillens (*Johannes*

Table B3: Reconstructed sizes of tiles, windows and overall room heights in eleven paintings (dimensions in centimetres)

Painting	*Tiles*			*Room height*
	Ceramic floor	*Marble floor*	*Delftware skirting*	
The Glass of Wine	14.6			
The Girl with a Wineglass	14.6			
Woman with a Lute		29.3		
Lady Writing a Letter, with her Maid		29.3	12.7	
The Music Lesson		29.3		322.6
The Concert		29.3		
Allegory of Painting		29.3		270.5
Allegory of the Faith		29.3		261.9
The Love Letter		29.3		
Lady Standing at the Virginals		29.3	13.5	
Lady Seated at the Virginals		29.3	13.5	
Averages			13.2	285.0
Actual sizes of surviving examples			13.0	

Vermeer, p. 78) mentions similar 17th-century tables in the Rijksmuseum in Amsterdam which are 75 cm high, 80 to 85 cm wide and 130 to 150 cm long. Clearly such tables were manufactured in a variety of sizes.

The tables in three more pictures are generally lower and narrower. However, these are all completely covered with carpets or cloths, and might be different items of furniture altogether.

I have argued that the virginals in the *Lady Standing* . . . and the *Lady Seated* . . . are one and the same instrument. They do indeed turn out to have very similar widths and overall heights. There is, however, a large disparity in their estimated lengths. Again, this is a difficult dimension to obtain with accuracy, because

Table B3 *continued:* Reconstructed sizes of tiles, windows and overall room heights in eleven paintings
(dimensions in centimetres)

Painting	*Windows*							
	Sill height	*Height to top of lower frame*	*Width of embrasures*		*Casement*		*Glass*	
			Far	*Near*	*Width*	*Height*	*Width*	*Height*
The Glass of Wine	86.4	177.8	132.1	144.8	63.5	78.7		
The Girl with a Wineglass	92.7	181.6			63.5	76.2		
Woman with a Lute	95.0	183.9					74.7	76.7
Lady Writing a Letter, with her Maid	92.7	182.9					58.4	72.4
The Music Lesson	102.1	227.1	173.7		78.7	88.9		
The Concert								
Allegory of Painting								
Allegory of the Faith								
The Love Letter								
Lady Standing at the Virginals	89.7	175.0					51.3	68.8
Lady Seated at the Virginals		190.2			54.9	?		
Averages	93.0	188.2	150.1		65.1	81.3	61.5	72.6

of the instruments being seen nearly end-on and from a low viewpoint. The virginals in *The Music Lesson* are obviously a different instrument altogether—indeed the dimensions are significantly larger than those of the virginals in the other two paintings. Table B4 below compares these instruments, and also the harpsichord in *The Concert*, with some actual contemporary Ruckers instruments. (The measurements here are for the cases alone, without the stands.) Vermeer's examples are all within or close to the size ranges of the real instruments, with the exception of the over-sized virginals in *The Music Lesson*.

Table B4: Actual and reconstructed sizes of keyboard instruments
(dimensions in centimetres: length × width × height)

	Actual size	*Size as estimated from reconstructions*
Ruckers harpsichords (without stands)	236.2 × 87.6 × 27.9 223.5 × 78.7 × 26.7 223.5 × 83.8 × 26.7	
Harpsichord in *The Concert*		218.4 × 81.3 × 25.9
Ruckers virginals (without stands)	166.4 × 47.7 × 24.1 142.2 × 48.8 × 21.6 171.4 × 49.5 × 24.1	
Virginals in *The Music Lesson*		195.1 × 56.4 × 22.5
Virginals in *Lady Standing at the Virginals*		154.9 × 52.3 × 27.5
Virginals in *Lady Seated at the Virginals*		207.8 × 47.0 × 22.5

The three Delftware jugs are all very similar in height (although the jug in *The Glass of Wine* has a slightly differing profile from those in the other two paintings).

One point to note in Table B2, as remarked in the text, is that not just the virginals but also the lions' head chairs in *The Music Lesson* seem to be somewhat bigger than in other pictures. A similar phenomenon is found in Table B3 which compares the calculated dimensions of the windows in the 11 paintings, and where again those for *The Music Lesson* stand out as being larger. It is not obvious why this painting should be anomalous in this respect.

If we set on one side for the moment the sizes for *The Music Lesson*, we find a reasonably close agreement throughout in the sill heights and in the heights to the tops of the lower window frames. In two cases, *The Glass of Wine* and *The Girl with a Wineglass*, it is possible to obtain good measurements of the casements, since in both paintings one casement—that with the stained glass centrepiece—is open and is seen more or less frontally. (In neither case is the left-hand edge visible, but the overall width can be confidently determined by reference to the pattern of leading.)

In other paintings some of the calculated window dimensions are more variable: but here the casements are closed and so are viewed at shallow angles. Sometimes they are partially hidden by curtains; and the nearer casement is always partly concealed because it is recessed in the window embrasure. For these reasons we might expect more consistency in the estimated heights than in the widths. This is indeed what is found, as for example in the heights of the glazed areas of the casements.

There are only three pictures in which the ceiling is visible, and so where the overall height of the room can be estimated. While these heights are almost identical for the *Allegory of Painting* and the *Allegory of the Faith*, the third measurement, for *The Music Lesson*, is again anomalously large.

On the website that accompanies this book (www.vermeerscamera.co.uk) there are reconstructed plans and side views for all 11 paintings, in the style of Figures 33 and 35, along with bird's-eye views like that shown for *The Music Lesson* in Figure 40. The website also contains detailed measured drawings of Vermeer's room and all the furniture from which the scale models were constructed.

NOTES

ACKNOWLEDGEMENTS

1. M J Kemp, *The Science of Art: Optical Themes in Western Art from Brunelleschi to Seurat*, Yale University Press, New Haven, Conn. (1990), pp. 195–6.

2. P Steadman, 'In the studio of Vermeer', in R Gregory, J Harris, P Heard, and D Rose (eds), *The Artful Eye*, Oxford University Press, Oxford (1995), pp. 353–72.

3. M P van Maarseveen, *Vermeer of Delft: His Life and Times*, Stedelijk Museum het Prinsenhof Delft and Bekking Publishers, Amersfoort (1996).

4. H L Houtzager, G C Klapwijk, H W van Leeuwen, M A Verschuyl, and W F Weve (eds), *De Kaart Figuratief van Delft*, Elmar, Rijswijk (1997).

INTRODUCTION

1. Kemp, *The Science of Art*, p. 196.

2. Ibid.

3. L Gowing, *Vermeer*, Faber, London (1952; 2nd edn 1970). See especially pp. 18–25.

4. Ibid., p. 23.

5. Ibid.

1 THE CAMERA OBSCURA

1. Much of the historical information in this chapter comes from four sources: J Waterhouse, 'Notes on the early history of the camera obscura', *The Photographic Journal* vol. XXV, 9 (31 May 1901), pp. 270–90; G Potonniée, *Histoire de la Découverte de la Photographie*, Paris (1925), trans. E Epstean as *The History of the Discovery of Photography*, Tennant and Ward, New York (1936), chapters III–V; H Gernsheim with A Gernsheim, *The History of Photography: From the Camera Obscura to the Beginning of the Modern Era*, Thames and Hudson, London (1955; rev. 1969), chapter 1; and J H Hammond, *The Camera Obscura: A Chronicle*, Adam Hilger, Bristol (1981).

2. The camera obscura is named as such in the index to Johannes Kepler, *Ad Vitellionem Paralipomena*, Frankfurt (1604) and on p. 16 of Johannes Kepler, *Dioptrice*, Augsburg (1611).

3. Aristotle, *Problems*, ed. W S Hett, Heinemann, London (1936), Book XV, p. 341.

4. J Needham, *Science and Civilisation in China*, Cambridge University Press, Cambridge (1962), vol. 4, part 1, pp. 97–9; Alhazen [Ibn al-Haitham], *Opticae Thesaurus Alhazeni Arabis*, ed. F Risner, Basel (1572; repr. New York 1972). According to D C Lindberg in 'The theory of pinhole images from antiquity to the thirteenth century' (*Archive for History of Exact Sciences* 5,1968–9, pp. 154–76), Alhazen was the first person to analyse the principles of the camera obscura correctly.

5. See Lindberg, 'The theory of pinhole images'.

6. John Pecham, *Perspectiva Communis*, part 1, prop. 5 (1279); English edn D C Lindberg (ed.), *John Pecham and the Science of Optics: Perspectiva Communis*, Madison, Wis. (1970). This translation is taken from Gernsheim, *The History of Photography*, p. 18.

7. Reinerus Gemma Frisius, *De Radio Astronomico et Geometrico*, Antwerp and Louvain (1545). Figure 1 is from the 2nd edn (1558), p. 39.

8. I A Richter (ed.), *Selections from the Notebooks of Leonardo da Vinci*, Oxford University Press, Oxford (1977), pp. 115–16.

9. R Corson, *Fashions in Eyeglasses*, Peter Owen, London (1967), p. 20. See also C Singer, E J Holmyard, A R Hall, and T I Williams (eds.), *A History of Technology*, Clarendon Press, Oxford (1957), vol. 3, p. 230.

10. Girolamo Cardano, *De Subtilitate*, Nuremburg (1550), book IV, p. 107. This translation is taken from Potonniée, *The History of the Discovery of Photography*, p. 14.

11. Gernsheim, *The History of Photography*, p. 20, and Potonniée, *The History of the Discovery of Photography*, p. 14.

12. Waterhouse, 'Notes on the early history of the camera obscura', p. 274.

13. Daniele Barbaro, *La Pratica della Perspettiva*, Borgominieri, Venice (1568).

14 Ibid., part 9, ch. 7, p. 193. This translation is taken from Waterhouse, 'Notes on the early history of the camera obscura', p. 276.

15. Giovanni Battista Benedetti, *Diversarum Speculationem Mathematicarum et Physicarum Liber*, Turin (1585).

16. Box-type cameras with 45 degree mirrors are described by Johann Christoph Sturm in *Collegium Experimentale, sive Curiosum*, Nuremburg (1676) and illustrated by Johann Zahn, *Oculus Artificialis Teledioptricus, sive Telescopium*, Wurzburg (1685–6). The viewing screens on instruments of this type were sometimes made of plain glass covered with oiled paper, so combining translucency with rigidity.

17. Giovanni Battista della Porta, *Magia Naturalis*, 1st edn Naples (4 vols, 1558; 2nd edn, 20 vols, 1589). The description of the camera in the 1st edn is in vol. IV, cap. ii, p. 143.

18. *Natural Magick by John Baptista Porta, a Neopolitane*, trans. Thomas Young and Samuel Speed, London (1658), p. 364.

19. Kepler, *Ad Vitellionem Paralipomena*, Frankfurt (1604). The camera is described in chapter 2, pp. 51–4. On p. 300 Kepler mentions observations of the sun made using a camera obscura in 1600. In Kepler's *Dioptrice* (Augsburg, 1611) Problem XLIII, p. 16 concerns the camera.

20. Christopher Scheiner, *Rosa Ursina sive Sol*, Bracciano, Rome (1626–30), book II, p. 107. The projecting telescope is illustrated in book II, p. 77.

21. Max Caspar, *Kepler*, Abelard-Schuman, London (1959), p. 252.

22. Hammond, *The Camera Obscura*, cites G Grigson and C H Gibbs-Smith, *People, Places and Things*, Waverley, London (1954), vol. 3.

23. Henry Wotton, *Reliquiae Wottonianae*, London (1651), p. 413.

24. The design of telescope which has become known as 'Keplerian' has two convex lenses. It is the 'Galilean' telescope which has one convex and one concave lens. However, Kepler described and experimented with both arrangements. Kepler did indeed design the 'Keplerian'; Galileo did not design the 'Galilean'.

25. Giovanni Francesco Costa, 'Veduta del Canale verso la Chiesa della Mira', from *Delle Delicie del Fiume Brenta*, Venice (1750–6).

26. Athanasius Kircher, *Ars Magna Lucis et Umbrae*, Rome (1646), plate 28. According to Gernsheim, *The History of Photography*, p. 23, Kircher saw this design of camera in Germany.

27. Hammond, *The Camera Obscura*, p. 26.

28. According to J-F Niceron, *La Perspective Curieuse*, Paris (3rd edn, 1652), there was a camera obscura set up as an entertainment at the Samaritaine on the Pont-Neuf in Paris, in the mid-17th century. However, it seems that this did not allow views of the surrounding townscape, but was used instead for a kind of magic show, by projecting images of marionettes, or of real people in woodland or garden settings, presumably by setting them against painted backgrounds.

29. Thomas Sandby, *Windsor from the Gossels*, collection of HM The Queen, Windsor Castle, 12.5 x 52.8 cm. The inscription reads 'drawn in a camera T. S.'

30. Kaspar Schott, *Magia Universalis Naturae et Artis*, Würzburg (1657), book IV, p. 200; Johann Christoph Sturm, *Collegium Experimentale, sive Curiosum*, Nuremburg (1678); Zahn, *Oculus Artificialis Teledioptricus*.

31. Editions of *Magia Naturalis* were published in Antwerp in 1564 and Leyden in 1651. See A K Wheelock, *Perspective, Optics and Delft Artists Around 1650*, Garland, New York (1977), pp. 177 and 178.

32. For biographical information about Constantijn Huygens relevant to the present discussion, see A G H Bachrach, 'The role of the Huygens family in seventeenth-century Dutch culture', in H J M Bos, M J S Rudwick, H A M Snelders, and R P W Visser (eds.), *Studies on Christiaan Huygens*, Swets and Zeitlinger, Lisse (1980), pp. 27–52; also R L Colie, *'Some Thankfulnesse to Constantine': A Study of English Influence upon the Early Works of Constantijn Huygens*, Nijhoff, The Hague (1956).

33. For biographical information on Cornelis Drebbel, see Colie, *'Some Thankfulnesse to Constantine'*, pp. 93–110, and L E Harris, *The Two Netherlanders: Humphrey Bradley and Cornelis Drebbel*, W Heffer and Sons, Cambridge (1961).

34. Bachrach, 'The role of the Huygens family', p. 37.

35. Ibid., p. 45; although Colie, *'Some Thankfulnesse to Constantine'*, p. 93, says that Drebbel married Goltzius's *daughter*.

36. Colie, *'Some Thankfulnesse to Constantine'*, p. 93.

37. Constantijn Huygens, *De Briefwisseling 1608–1687*, ed. J A Worp, Nijhoff, 's Gravenhage (1911–16), vol. I, p. 89, quoted in Colie, *'Some Thankfulnesse to Constantine'*, p. 106.

38. Huygens, *De Briefwisseling 1608–1687*, vol. I, p. 94, quoted in Colie, *'Some Thankfulnesse to Constantine'*, p. 106.

39. 'Fragment eener autobiographie van Constantijn Huygens', ed. J A Worp, in *Bijdragen en Mededeelingen der Historisch Genootschap* XVIII (1897); quoted by Colie, *'Some Thankfulnesse to Constantine'*, p. 106.

40. Bachrach, 'The role of the Huygens family', p. 41

41. A H Kan (ed.), *De Jeugd van Constantijn Huygens door Hemzelf Beschreven*, Donker, Rotterdam (1946), appendix, pp. 137–40, cited in Colie, *'Some Thankfulnesse to Constantine'*, p. 107.

42. See Erik Barnouw, 'Torrentius and his camera', *Studies in Visual Communication*, vol. 10, no. 3, summer 1984, pp. 22–9. Barnouw in turn draws on Abraham Bredius, *Johannes Torrentius, Schilder, 1589–1644*, Nijhoff, The Hague (1909).

43. *De Jeugd van Constantijn Huygens*; quoted by Barnouw, 'Torrentius and his camera', p. 22.

44. Johannes Torrentius [Johan van der Beeck], *Still life*, 1614, Rijksmuseum, Amsterdam. In 1913 the painting was rediscovered in a Dutch grocery shop, where it was being used as the lid for a barrel of currants. The picture does not, on close inspection, show any of the tell-tale signs of a camera technique (as discussed in the next chapter)—with perhaps one exception. The goblet at the centre is not as sharply and crisply rendered as glassware in some other Dutch still-life paintings, but has a slightly matt quality—as if seen through a ground glass screen.

45. See Colie, *'Some Thankfulnesse to Constantine'*, p. 107.

46. C Brown, *Carel Fabritius*, Phaidon, Oxford (1981).

47. G Q Williams and P Kemp, *The Music Dealer: An Explanation of Distorted Perspective*, *http://gate.uwe.ac.uk:8000/~fj-maddi/qw/musicdeal.html*. Other discussions of the painting can be found in A K Wheelock, 'Carel Fabritius: perspective and optics in Delft', *Nederlands Kunsthistorische Jaarboek* 1973, pp. 63–83; and W A Liedtke, '"The View in Delft" by Carel Fabritius', *The Burlingon Magazine* no. 875, vol. 118 (1976), pp. 61–73.

48. Susan Koslow, 'De wonderlijke perspectyfkas: an aspect of seventeenth-century Dutch painting', *Oud Holland* 82 (1967), pp. 35–56. See also C Brusati, *Artifice and Illusion: The Art and Writing of Samuel Van Hoogstraten*, Chicago University Press, Chicago, Ill. (1995), pp. 169–217.

49. Arthur Wheelock says, in more than one place, that van Hoogstraten actually *built* two cameras himself: see, for example, A K Wheelock, 'Vermeer of Delft: his life and his artistry', in A K Wheelock (ed.), *Johannes Vermeer*, catalogue of an exhibition held at the National Gallery of Art, Washington and the Mauritshuis, The Hague, Yale University Press, New Haven, Conn. (1995–6), pp. 15–29; esp. p. 26. But the evidence from the *Inleyding* (see n. 50 below) suggests rather that he was privileged only to experiment (in London and Vienna) with instruments constructed by others.

50. Samuel van Hoogstraten, *Inleyding tot de Hooge Schoole der Schilderkonst: Anders de Zichtbaere Werelt . . .*, Rotterdam (1678).

51. A K Wheelock, *Vermeer and the Art of Painting*, Yale University Press, New Haven, Conn. (1995), pp. 18–19. Also J-L Delsaute, 'The camera obscura and painting in the sixteenth and seventeenth centuries', in I Gaskell and M Jonker (eds.), *Vermeer Studies*, National Gallery of Art, Washington and Yale University Press, New Haven, Conn. (1998), pp. 111–23; see esp. p. 113, where Delsaute claims that van Hoogstraten 'avoids describing any modus operandi that would definitely allow the painter to use the camera obscura for drawing.'

52. Van Hoogstraten, *Inleyding*, p. 263. This translation from Brusati, *Artifice and Illusion*, p. 71.

53. Van Hoogstraten, *Inleyding*, p. 263.

54. Ibid. Brusati (*Artifice and Illusion*, p. 286 n. 44) also mentions a camera constructed in Dordrecht by van Hoogstraten's compatriot Dr Johan van Beverwijck. It is illustrated in a treatise published in 1642 that contains van Hoogstraten's earliest dated print.

55. Brusati, *Artifice and Illusion*, p. 54.

56. Ibid., p. 285 n. 38.

57. Ibid., p. 70.

58. Ibid., pp. 72–3 and Figure 43.

59. See P T A Swillens, *Johannes Vermeer: Painter of Delft 1632–1675*, Spectrum, Utrecht and Brussels (1950), p. 187.

60. William Hyde Wollaston took out a patent in 1806. See 'Description of the camera lucida, communicated by the author', *The Philosophical Magazine*, 1st series, XXVII, London (1807), pp. 343–7. Also J-B Amici, 'Mémoire de M. le Professeur J-B Amici de Modène, sur les chambres claires de son invention', *Annales de Chimie et de Physique* XXII (1823), pp. 137–55.

61. H Schwarz, 'Vermeer and the camera obscura', *Pantheon* XXIV (May–June 1966), pp. 170–82, citing E Hoppe, *Geschichte der Optik*, Leipzig (1926).

2 THE DISCOVERY OF VERMEER'S USE OF THE CAMERA

1. J Reynolds, *The Literary Works of Sir Joshua Reynolds*, Cadell, London and Blackwood, Edinburgh (1835), vol. II, p. 204.

2. For an account of the painting's acquisition by the Mauritshuis, see Wheelock (ed.), *Johannes Vermeer*, catalogue entry on *View of Delft*, pp. 120–7.

3. William Bürger [Théophile Thoré], 'Van der Meer de Delft', *Gazette des Beaux-Arts* 21 (1866), pp. 297–330, 458–70, 542–75.

4. See, for example, the perspective construction drawings by the architectural painter Pieter Saenredam, reproduced in *Dutch Church Painters: Saenredam's 'Great Church at Haarlem' in Context*, catalogue of an exhibition at the National Gallery of Scotland,

Edinburgh, 6 July–9 September 1984. See also the discussion in Kemp, *The Science of Art*, pp. 114–16.

5. The most exhaustive collection and analysis of documents relating to Vermeer's life is in J M Montias, *Vermeer and his Milieu: A Web of Social History*, Princeton University Press, Princeton, NJ (1989). See Appendix B, pp. 268–368.

6. Ibid., pp. 339–44.

7. Montias (ibid., pp. 103–7) discusses a series of possible candidates for Vermeer's master, but his efforts to find out more about the painter's apprenticeship have been, in his own words, 'largely unsuccessful'.

8. W J 's Gravesande, *Essai de Perspective*, Paris (1711). This translation is taken from A Hyatt Mayor, 'The photographic eye', *Bulletin of the Metropolitan Museum of Art*, new series vol. V, no. 1 (1946), pp. 15–26. See p. 20.

9. Count F Algarotti, *Saggio sopra la Pittura*, Livorno (1763). English translation, Glasgow (1764).

10. Edmond and Jules Goncourt, *Journal*, 11 September 1861; Laffont, Paris (1989), vol. 1, p. 727. 'Une RUE DE DELFT: le seul maître qui ait fait de la maison briquée du pays un daguerréotype animé par l'esprit.'

11. J Pennell, 'Photography as a hindrance and a help to art', *British Journal of Photography* no. 1618, vol. XXXVIII (1891), pp. 294–6.

12. Ibid., p. 295.

13. Arthur Wheelock argues that the perspective of *Officer and Laughing Girl* has *two* horizon lines—a geometrical inconsistency—as a result of Vermeer making the perspective of the cavalier's chair converge to a higher vanishing point than that of the rest of the scene (*Vermeer*, Thames and Hudson, London, 1981, p. 82). It is a little difficult to determine the exact angles of the converging lines in the chair in question: arguably the perspective is geometrically sound.

14. For a fuller explanation see chapter 8, p. 149 and n. 30.

15. R H Wilenski, *An Introduction to Dutch Art*, Faber and Gwyer, London (1928), pp. 280–90.

16. Ibid., p. 284.

17. Gowing, *Vermeer*, p. 125, n. 79.

18. Hyatt Mayor, 'The photographic eye'. Hyatt Mayor refers (p. 19) to a suggestion by one Allyn Cox that Vermeer might have used the camera, but I have failed to track this down.

19. Ibid., p. 20.

20. D A Fink, 'Vermeer's use of the camera obscura—a comparative study', *Art Bulletin* 53 (1971), pp. 493–505. See p. 496.

21. C Seymour Jr, 'Dark chamber and light-filled room: Vermeer and the camera obscura', *Art Bulletin* 46 (1964), pp. 323–31.

22. Ibid., p. 325.

23. Ibid., pp. 325–6.

24. Ibid., p. 327.

25. The panels measure 23 x 18 cm (*Girl with a Red Hat*) and 20 x 18 cm (*Young Girl with a Flute*). One problem for the hypothesis of a box camera on a table—which Seymour overlooks—is presented by the position and orientation of the lions' heads in *Girl with a Red Hat*. The girl is not sitting on the chair in question: the lions' heads face *towards* us. She is *behind* the chair. Comparison with other pictures where the same chairs appear—for example, *Officer and Laughing Girl*—show that the lions' heads would normally be at the level of a sitting woman's shoulders, not her forearm. Either the red-hatted girl is too high up or the chair back is too low. Perhaps she is standing? Wheelock (*Vermeer*, p. 130) has also pointed out that the representations of the two lions' heads are not far enough apart to correspond to the chairs which they appear on elsewhere. In later chapters I will present measurements and drawings which confirm this observation. None of this necessarily contradicts the evidence of the lack of focus, however, or the possibility that a box-type camera—perhaps supported nearer to the eye level of a standing person—might have been used. Images of one or both lions' heads might have been traced, moved, and painted in altered positions.

26. Swillens, *Johannes Vermeer*, p. 92 and plate 59b.

27. Gowing, *Vermeer*, p. 128. Gowing says that 'The landscapes [*The Little Street* as well as *View of Delft*] were apparently painted from first floor windows.' This is certainly true of *The Little Street*, but Swillens's house on Blaeu's plan—if the detail is to be believed—looks to be single-storey. (On the other hand there are two-storey houses in this position in the 1675–8 Pictorial Map.)

28. Wheelock, *Vermeer*, p. 96.

29. Wheelock, *Perspective, Optics and Delft Artists Around 1650*, p. 294.

30. Gowing, *Vermeer*, p. 128.

31. Fink, 'Vermeer's use of the camera obscura'.

32. Wheelock, *Perspective, Optics and Delft Artists Around 1650*, p. 287.

33. Ibid., p. 289.

34. One such appears in Gabriel Metsu, *Lady Reading a Letter*, c. 1666–7, where the lady's maid is pulling back the curtain on a seascape.

35. Gerrit Dou, *Self Portrait*, c.1645, Rijksmuseum, Amsterdam. Another example is Gerrit Houckgeest, *Interior of the Old Church in Delft*, c.1650–1, Rijksmuseum, Amsterdam.

36. J A Welu, 'Vermeer: his cartographic sources', *Art Bulletin* 57, 1975, pp. 529–47.

37. Ibid., p. 531.

38. Fink ('Vermeer's use of the camera obscura', p. 501) claims that the Blaeu map is seen not frontally but at a very slight angle, and for this reason the top and bottom edges converge very slightly towards the right in Vermeer's reproduction. I do not find this phenomenon, nor does Wheelock. Welu ('Vermeer: his cartographic sources', p. 531) suggests that the map may appear in Vermeer's picture somewhat smaller (to scale) than its actual size. The question is difficult to settle in this particular case, but the calculations in Chapter 5 indicate, to the contrary, that Vermeer reproduces the maps in his pictures at their precise known sizes.

39. See Wheelock, *Perspective, Optics and Delft Artists Around 1650*, ch. 7, pp. 261–321; Wheelock, *Vermeer* p. 39; Wheelock, 'Vermeer of Delft: his life and artistry', pp. 25–7; Wheelock, *Vermeer and the Art of Painting*, pp. 17–19 and *passim*.

40. Wheelock, *Perspective, Optics and Delft Artists Around 1650*, p. 298.

41. Ibid., p. 276–7.

42. Ibid., pp. 297–8.

43. Ibid., p. 299.

44. Gowing, *Vermeer*, p. 20.

45. Ibid., p. 137.

46. K M Groen, I D van der Werf, K J van den Berg, and J J Boon, 'Scientific examination of Vermeer's *Girl with a Pearl Earring*', in I Gaskell and M Jonker (eds.), *Vermeer Studies*, pp. 168–83. See pp. 170–1.

47. Gowing, *Vermeer*, pp. 137–8.

48. Ibid., p. 23.

3 WHO TAUGHT VERMEER ABOUT OPTICS

1. For biographies of Leeuwenhoek, see C Dobell, *Antony van Leeuwenhoek and his 'Little Animals'*, John Bale, Sons and Danielsson, London (1932) and A Schierbeek, *Measuring the Invisible World: The Life and Works of Antoni van Leeuwenhoek FRS*, Abelard-Schumann, London (1959).

2. See R S Clay and T H Court, *The History of the Microscope*, Charles Griffin, London (1932), pp. 8–10.

3. Dobell, *Antony van Leeuwenhoek and his 'Little Animals'*, p. 362.

4. Ibid., p. 317.

5. See Schierbeek, *Measuring the Invisible World*, pp. 86 ff.

6. H G Plimmer, 'Bedellus immortalis', *Journal of the Royal Microscopical Society*, April 1913, pp. 120–35; see esp. p. 127. Also Dobell, *Antony van Leeuwenhoek and his 'Little Animals'*, pp. 40 and 42. Graaf was the discoverer of the 'Graafian follicle', which carries the ovum in the mammalian ovary. Huygens and Leeuwenhoek corresponded on microscopical matters, although from what date is unknown. Certainly by the 1670s they were in regular contact. (See E G Ruestow, *The Microscope in the Dutch Republic: The Shaping of Discovery*, Cambridge University Press, Cambridge, 1996, p. 149.)

7. See Dobell, *Antony van Leeuwenhoek and his 'Little Animals'*, pp. 392–7 and Schierbeek, *Measuring the Invisible World*, pp. 205–9. Leeuwenhoek was thus by no means the first to publish discoveries with the microscope. As early as 1592 there was a work on insects by George Huefnagel printed in Frankfurt, illustrated by drawings made with a microscope—most probably a single lens. Other similar books appeared from the 1620s. In 1656 Pierre Borel published his *Century of Microscopical Observations*. Robert Hooke's famous and influential *Micrographia* followed soon after in 1665.

8. Dobell, *Antony van Leeuwenhoek and his 'Little Animals'*, p. 29.

9. Van Maarseveen, *Vermeer of Delft*, p. 89.

10. See Montias, *Vermeer and his Milieu*, pp. 224–5.

11. Dobell, *Antony van Leeuwenhoek and his 'Little Animals'*, pp. 31–3, and Schierbeek *Measuring the Invisible World*, pp. 19–21.

12. F D O Obreen, *Archief voor Nederlandsche Kunstgeschiedenis*, Van Hengel en Eeltjes, Rotterdam (1877–90): 'Iets over den Delftschen schilder Johannes Vermeer', vol. IV (1881), p. 289.

13. Dobell, *Antony van Leeuwenhoek and his 'Little Animals'*, p. 36.

14. Wheelock, *Vermeer*, p. 13.

15. Wheelock, *Perspective, Optics and Delft Artists Around 1650*, p. 285.

16. Montias, *Vermeer and his Milieu*, pp. 225–6.

17. Ibid., p. 229.

18. J M Montias, 'Recent archival research on Vermeer', in I Gaskell and M Jonker (eds), *Vermeer Studies*, pp. 93–109. See p. 102.

19. Dobell, *Antony van Leeuwenhoek and his 'Little Animals'*, pp. 342–5.

20. R Boitet (ed.), *Beschryving der Stadt Delft* (1729); quoted by Schierbeek, *Measuring the Invisible World*, pp. 23–4.

21. Dobell, *Antony van Leeuwenhoek and his 'Little Animals'*, p. 51.

22. Schierbeek, *Measuring the Invisible World*, p. 17.

23. A Kircher, *Ars Magna Lucis et Umbrae.*

24. C Reilly, *Athanasius Kircher SJ: Master of a Hundred Arts, 1602–1680*, Edizioni del Mondo, Wiesbaden/Rome (1974), p. 86.

25. A Kircher, *Scrutinium Physico-Medicum Contagiosae Luis, Quae Pestis Dicitur*, Rome (1658).

26. Dobell, *Antony van Leeuwenhoek and his 'Little Animals'*, pp. 365–7.

27. Ibid., pp. 161–2, and Schierbeek, *Measuring the Invisible World*, p. 67.

28. William Eamon lists all the languages in which della Porta's *Magia Naturalis* was published, including Dutch, in *Science and the Secrets of Nature: Books of Secrets in Medieval and Early Modern Culture*, Princeton University Press, Princeton, NJ (1994), p. 206.

29. Schwarz ('Vermeer and the camera obscura', p. 177) suggests that Leeuwenhoek might also have been familiar with Jean-François Niceron's *Thaumaturgus Opticus . . . ad Cardinalem Mazarinum*, which illustrates the camera obscura, since Niceron was himself a microscopist. Yet another possibility is Christopher Scheiner's *Rosa Ursina sive Sol* (see Chapter 1 and Figure 5).

30. Schwarz, 'Vermeer and the camera obscura', p. 174. Leeuwenhoek's letter to Oldenbourg is dated 26 March 1675. Jørgen Leisner has questioned Schwarz's interpretation, however (personal communication). As Leisner reads the passage, it is his method of observation using the microscope that Leeuwenhoek wants to keep secret, rather than his method of drawing.

31. E V Lucas, *Vermeer of Delft*, Methuen, London (1922).

32. Now in the Rijksmuseum, Amsterdam. Some other portraits of Leeuwenhoek, including one on his signet ring, seem to derive from Verkolje's. There are two exceptions. A small oval engraving by Jan Goeree, showing Leeuwenhoek aged 75, appears on the title page of the fourth volume of the collected *Brieven* [*Letters*], Delft (1718). Leeuwenhoek also figures in a group portrait by Cornelis de Man, 'The Anatomy Lesson of Cornelis 's Gravesande', dated 1681, now in the Prinsenhof Museum in Delft (see van Maarseveen, *Vermeer of Delft*, p. 88). Dr 's Gravesande, who is demonstrating on the cadaver, was Leeuwenhoek's neighbour and taught him some anatomy. Leeuwenhoek himself is immediately behind and to the right of the doctor, with his hand on his chest.

33. A copy is also in the Rijksmuseum, Amsterdam.

34. Welu, 'Vermeer: his cartographic sources', pp. 544–6. See in particular n. 87 p. 545.

35. A Blankert (*Vermeer of Delft*, Phaidon, Oxford, 1978, p. 166) has questioned whether the date on *The Geographer* is not a later addition.

36. Dobell, *Antony van Leeuwenhoek and his 'Little Animals'*, p. 353.

37. Wheelock, *Vermeer*, p. 13, and Wheelock (ed.), *Johannes Vermeer*, p. 172.

38. Montias, *Vermeer and his Milieu*, p. 225.

39. Dobell, *Antony van Leeuwenhoek and his 'Little Animals'*, p. 354.

40. Constantijn Huygens, 'Fragment eener autobiographie van Constantijn Huygens', pp. 100–4. See also A E Bell, *Christian Huygens and the Development of Science in the Seventeenth Century*, Edward Arnold, London (1947), p. 17.

41. Bachrach, 'The role of the Huygens family', pp. 38–9. Also Colie, *'Some Thankfulnesse to Constantine'*, ch. VII.

42. Bell, *Christian Huygens and the Development of Science*, p. 17.

43. R Descartes, *Dioptrique*, Leyden (1637). See also Clay and Court, *The History of the Microscope*, p. 14. Descartes's illustrations were copied by Athanasius Kircher in *Ars Magna Lucis et Umbrae*.

44. Clay and Court, *The History of the Microscope*, p. 8.

45. Ibid., pp. 8–9.

46. Ibid., p. 12.

47. Giovanni du Pont, 'Relazioni dei Viaggi', manuscript in National Library, Paris (Fonds Périgord, t. VI Cart. 20 *et seq*.), quoted in Clay and Court, *The History of the Microscope*, p. 11.

48. Clay and Court, *The History of the Microscope*, p. 13.

49. H J M Bos, M J S Rudwick, H A M Snelders, and R P W Visser (eds.), *Studies on Christiaan Huygens*, Swets and Zeitlinger B V, Lisse (1980), introduction, p. 11.

50. Bell, *Christian Huygens and the Development of Science*, p. 26.

51. B Broos, 'Un celebre Peijntre nommé Verme[e]r', in Wheelock (ed.), *Johannes Vermeer*, pp. 47–65. See esp. pp. 49–50.

52. Manuscript in Koninklijke Bibliothek, The Hague, inv. no. 129 D 16, vol. 1.

53. Broos, 'Un celebre Peijntre nommé Verme[e]r', p. 51. Broos also mentions that the English sculptor Johan (or Jean) Larson, another friend of the Huygens family, bought a *tronie* [portrait head] from Vermeer, probably in 1660. He also draws attention to the fact that Constantijn Huygens ordered a virginals in 1648 from the Ruckers workshop, very similar to that which appears in *The Music Lesson*. 'Is it not possible', he asks, 'that Vermeer saw the "Ruckers" in the Huygens residence?'

54. Montias, 'Recent archival research on Vermeer'. See pp. 99–100.

55. Balthasar de Monconys, *Journal des Voyages de Monsieur de Monconys*, Boissat et Remeus, Lyons (1665), part II, p. 133.

56. Ibid., p. 149.

57. Broos, 'Un celebre Peijntre nommé Verme[e]r', p. 49.

58. Schwarz, 'Vermeer and the camera obscura', pp. 170–82.

59. Monconys, *Journal des Voyages*, part II, p. 145.

60. Ibid., pp. 153 and 161. See also Ruestow, *The Microscope in the Dutch Republic*, p. 22, and C. Wilson, *The Invisible World: Early Modern Philosophy and the Invention of the Microscope*, Princeton University Press, Princeton, NJ (1995), pp. 77 and 79.

61. Clay and Court, *The History of the Microscope*, pp. 21–2, citing Monconys, *Journal des Voyages*, p. 128.

62. Cited in Clay and Court, *The History of the Microscope*, p. 22.

63. Monconys, *Journal des Voyages*, part II, pp. 444 and 452.

64. Ibid., pp. 28 ff.

65. Ibid., p. 40.

66. Ibid., pp. 17–18.

67. Quoted by Bell, *Christian Huygens and the Development of Science*, p. 53.

68. Robert Boyle, 'Of the Systematicall or Cosmical Qualities of Things' , p. 18, in *Tracts*, Davis, Oxford (1671).

69. Schwarz, 'Vermeer and the camera obscura', p. 173.

4 A ROOM IN VERMEER'S HOUSE?

1. P T A Swillens, 'Een perspectivische studie over de schilderijen van Johannes Vermeer van Delft', *Oude Kunst* VII, 1929–30, pp. 129–61, and P T A Swillens, *Johannes Vermeer*.

2. Swillens, *Johannes Vermeer*, pp. 69–77 and plates 43–53.

3. Ibid., pp. 24 and 95, largely on the assumption that *The Little Street* was painted from a rear window of the inn.

4. Reproduced in facsimile in Houtzager et al. (eds.), *De Kaart Figuratief van Delft*. See sheets 45 (p. 131) and 46 (p. 133).

5. Abraham Rademaker and Leonard Schenk, *View of the Market*, *c.* 1720, both in the Delft Municipal Archive.

6. The earliest evidence of Vermeer sharing the house of Maria Thins is when a child of his, recorded as 'living on the Oude Langedijck', is buried in the Oude Kerk in 1660.

7. See Montias, *Vermeer and his Milieu*, p. 132.

8. Ibid., p. 341.

9. Swillens, *Johannes Vermeer*, p. 24 and plate 35c.

10. Montias, *Vermeer and his Milieu*, p. 176 n. 20 and Figure 31. In J M Montias, 'A chronicle of a Delft family', in A Blankert, J M Montias, and G Aillaud, *Vermeer*, Rizzoli, New York (1988), Figure 9, p. 23, Maria Thins's house is marked in a different position, but Professor Montias says that this is incorrect and the house indicated in my Figure 20 is the one he believes to have belonged to Maria Thins (personal communication). Montias says (*Vermeer and his Milieu*, p. 176) that 'If the room in which Vermeer painted . . . had three windows . . . it could only have faced west [i.e. onto the Molenpoort].' This seems incorrect, for reasons explained in Chapter 5 n. 31. Montias says that we do not know how accurately the houses in the *Figurative Map* were drawn. This question is, however, discussed in some detail in Houtzager et al., *De Kaart Figuratief van Delft*, pp. 23–36. Comparison with modern maps shows that the *Figurative Map* is not absolutely reliable, especially in the positions of some minor streets and individual buildings, but the surveying is done with care, and in general the Map is extremely informative and detailed for its period.

11. Hofstede de Groot writes, in the late 19th century, of a part of a mantelpiece being visible on the right of *Girl Interrupted at Her Music*, but it is not there today ('De een en dertigste Vermeer', *De Nederlandsche Spectator* 1899, p. 224).

12. *The Concert*, *Allegory of Painting*, *The Love Letter*, *Lady Writing a Letter, with her Maid*, *Lady Standing at the Virginals*, and *Lady Seated at the Virginals*.

13. *The Milkmaid*, *Lady Writing a Letter, with her Maid*, *The Geographer*, *Lady Standing at the Virginals*, and *Lady Seated at the Virginals*.

14. *Woman Holding a Balance*, *Lady Seated at the Virginals*, and *The Glass of Wine*.

15. Blankert (*Vermeer of Delft*, p. 159) says that the seated man in *The Girl with the Wineglass* was overpainted in the 18th century. The window in question, had it existed, would have been just to the left of this figure. On p. 171 Blankert says that the condition of *Girl Interrupted at her Music* is so poor, he had wondered whether it was 'the ruin of an original or a copy'. He quotes Hofstede de Groot's descriptions in 1899 ('De een en dertigste Vermeer') of damage to this picture from both extensive overpainting and vigorous cleaning. The birdcage, precisely in front of the position of a second possible window, is not by Vermeer but is a later addition.

16. Identified by E Neurdenburg as the coat of arms of the family of Janettje Jacobsdr. Vogel, the first wife of Vermeer's neighbour Moses J Nederveen, in 'Johannes Vermeer, eenige opmerkingen naar aanleiding van de nieuwste studies over den Delftschen schilder', *Oud Holland* 59 (1942), p. 69.

17. Swillens, *Johannes Vermeer*, pp. 73–4.

18. Ibid., pp. 70–3.

5 RECONSTRUCTING THE SPACES IN VERMEER'S PAINTINGS

1. As a matter of fact Leonardo himself gives a detailed procedure for doing just this (see K H Veltman, *Studies on Leonardo da Vinci: I Linear Perspective and the Visual Dimensions of Science and Art*, Deutsche Kunstverlag, Munich (1986), pp. 60–1 and 84–5). As he describes it: 'From a plane [i.e. a foreshortened perspective image of a flat gridded surface] on which a point is made by chance, to know how to test how distant it is and how far away one was standing to see this plane and how high the eye [was] . . .' F Dubery and J Willats, *Perspective and Other Drawing Systems* (Herbert Press, London, 1983, pp. 74–9) use Leonardo's method to reconstruct the space of Vermeer's *The Music Lesson*, and set up an axonometric [bird's-eye] view of the room seen from above, similar to my Figure 40.

2. W H Ittelson, *The Ames Demonstrations in Perception: A Guide to their Construction and Use*, Princeton University Press, Princeton, NJ (1952).

3. Difficulties arise in two circumstances. The first is where the principal surfaces of objects are seen at very shallow angles. The second is where a piece of furniture is close to the viewpoint and the positions of its feet cannot be seen. Both these problems are discussed further in Appendix B. Neither is very serious.

4. The analyses were made using tracings from reproductions of the pictures, typically around half-scale reductions of the actual works.

5. The fact that there are *two* distance points gives a double check, in the process of reconstruction, on this critical dimension. I have taken an average of the two measurements in every case.

6. Swillens, *Johannes Vermeer*, pp. 69–77 and plates 43–53.

7. In Swillens's plan the angle of view, in that part of the scene which is beyond the doorway, is 17°. In my plan it is nearer 12°.

8. B Nicolson, *Vermeer's 'Lady at the Virginals'*, Gallery Books no. 12, Lund Humphries, London (n. d.).

9. Gowing, *Vermeer*, p. 121.

10. Zahn, *Oculus Artificialis Teledioptricus*.

11. Swillens, *Johannes Vermeer*, p. 128.

12. Wheelock, *Vermeer*, p. 100.

13. Swillens (*Johannes Vermeer*, plate 65) reproduces Michiel van Musscher (?), *Artist's Studio*, Northbrook Collection, London. Two more boxes can be seen in Gonzales Coques, *Painter in his Studio*, Staatliches Museum, Schwerin, and an engraving by Vincent van der Vinne, *Painter in his Studio*, Rijksprentenkabinet, Amsterdam (see Wheelock (ed.), *Johannes Vermeer*, p. 74 Figure 12 and p. 76 Figure 14). There is also a 17th-century painter's box in the collection of the Rijksmuseum in Amsterdam, decorated with pastoral scenes attributed to Anthonie van Croos and Jan Martszen II—although it is much smaller than these other examples.

14. *The Glass of Wine*, *The Girl with a Wineglass*, *Woman with a Lute*, *The Concert*, and *Allegory of Painting*.

15. Swillens, *Johannes Vermeer*, plate 54a.

16. One copy of the Hondius original is preserved in the Library of the University of Amsterdam. The copy reproduced in Figure 43 is from the *Klencke Atlas* in the British Library in London. See Welu, 'Vermeer: his cartographic sources', pp. 535–6 and Figure 8 p. 537.

17. The surviving copy, still in loose sheets, is in the Bibliothèque Nationale in Paris. See Welu, 'Vermeer: his cartographic sources', pp. 536–8 and Figure 10 p. 537.

18. The map has been identified by Welu ('Vermeer: his cartographic sources', p. 533) as the same map of Holland and West Friesland by van Berckenrode (see Figure 15) which is seen in *Officer and Laughing Girl* and *Woman in Blue Reading a Letter*.

19. Dirck van Baburen, *The Procuress*, 1622, 101 x 107 cm, Museum of Fine Arts, Boston.

20. See S Sadie (ed.), *New Grove Dictionary of Music*, Macmillan, London (1980), vol. 19, pp. 791–2.

21. See Nicolson, *Vermeer's 'Lady at the Virginals'*, p. 4. Swillens, *Johannes Vermeer,* plate 55a illustrates another very comparable instrument (a similar case, if not a similar stand), dating from 1640, from the collection of the Rijksmuseum in Amsterdam. This is decorated with the same patterns of seahorses and has an equivalent printed motto on the lid.

22. As given, for example, in D Boalch, *Makers of the Harpsichord and Clavichord 1440–1840,* George Ronald, London (1956) and R Russell, *The Harpsichord and Clavichord,* Faber, London (1959).

23. See C H de Jonge, *Dutch Tiles*, Pall Mall Press, London (1971), p. 30, and E Neurdenburg and B Rackham, *Old Dutch Pottery and Tiles*, Benn, London (1923), p. 38. It is still possible to buy 17th-century examples in Delft's antique shops, at reasonable prices.

24. Swillens (*Johannes Vermeer,* p. 75) estimates the edge length of the marble tiles as 27 cm. Gowing (*Vermeer,* p. 125 n. 79) estimates the diagonal measurement of the tiles as between 15″ and 16″, which gives an edge length of 11″ (28 cm).

25. Vermeer's version of van Baburen's *Procuress* in *Lady Seated at the Virginals* does not, however, preserve the size, or even the proportions, of the actual painting, being somewhat stretched relative to the original. This phenomenon is discussed further in Chapter 7.

26. Swillens, *Johannes Vermeer,* p. 79 and plate 54c.

27. I am grateful to the museum authorities for locating the chairs and to Dr Marc van Leusen for measuring them.

28. See, for example, R Baarsen, *Dutch Furniture 1600–1800*, Rijksmuseum, Amsterdam/ Waanders Uitgevers, Zwolle (1993), p. 7, which gives overall dimensions for a purplewood chair with lions' heads in the Rijksmuseum collection, very similar to Vermeer's, as 105 cm (height), 44.5 cm (width), and 41.1 cm (depth). Marc van Leusen's measurements of similar chairs in the Prinsenhof Museum give dimensions of 105 cm (height), 43 cm (width), and 39 cm (depth). These compare with average values in the reconstructions of 107.6 × 42.4 × 42.4 cm.

29. Information from the national Dutch Building Agency and the City of Delft. Measurements of bricks in 17th-century houses around the Market Square in Delft give

typical lengths of 16 or 17 cm. I am grateful to Marc van Leusen for help on this question.

30. The 1649 map seems to be quite schematic especially in its rendering of houses, many of which are drawn in an identical format. Houtzager et al. (*De Kaart Figuratief van Delft*) show that the 1675–8 map was based on relatively accurate survey drawings, although the ground plan was then covered up, to an extent, by the bird's-eye views of buildings.

31. Montias (*Vermeer and his Milieu*, p. 176, n. 20) says that 'If the room in which Vermeer painted really had three windows as Swillens claims . . . it could only have faced west.' I believe that Montias has been misled here by the steep perspective of the pictorial maps into thinking that Maria Thins's house had a narrow frontage, insufficient for three windows facing north. The 1830 map shows, to the contrary, that this frontage—assuming it is correctly identified—was unusually wide. A room facing west onto the Molenpoort, even if lit by three windows, would not have been suitable as a painting studio, since this is only a narrow alley.

6 THE RIDDLE OF THE SPHINX OF DELFT

1. In the original calculations the 'projected image' on the back wall for *The Glass of Wine* worked out slightly smaller than the painting itself, and the 'projected image' for *The Girl with a Wineglass* was somewhat larger than the painting. These results formed the basis of an early diagram reproduced in Kemp (*The Science of Art*, p. 195) which differs for this reason from Figure 49 here. However, when these estimated viewpoints for the two paintings were used in the process of photographic reconstruction described in Chapter 7, it was found that the grids of floor tiles were *not* thereby reproduced with perfect accuracy. Adjustments were made experimentally to the viewpoints, and it emerged, by taking slightly changed positions, that the correct perspective geometry *was* precisely compatible in both cases with 'projected images' having the exact sizes of the actual paintings, on the back wall. The assumption about the sizes of the ceramic tiles remained unchanged. Figures 49 and 50 thus show—for simplicity—a composite of results from the photographic work for these two paintings, with the geometrical calculations for the remaining four.

2. $1/f = 1/x + 1/y$ where f is the focal length, x is the distance of lens to image plane (i.e. the back wall), and y is the distance of lens to subject. In this case x is typically around 90 cm (varying somewhat between paintings) and $y = 400$ cm to the nearer edge of the far window. This gives $f = 73$ cm.

3. Note further that the coincidence of the projected images is not, in itself, even dependent on the assumptions about absolute scale made in the last chapter—only on the assumptions that the marble tiles are *of equal size* in all relevant pictures, and that the ceramic tiles are half the size of the marble tiles. What precise values the tile dimensions are assumed to have has no effect on the geometry, other than to scale the entire plans and cross-sections of all the rooms up or down. It is only with the precise scale determined in the last chapter that the rectangles both possess the true sizes of the paintings and coincide *at the back wall.* At other scales they would coincide in some other plane, either further forward or further back: but they would still coincide. (On the other hand the maps and identifiable pieces of furniture would not then be at their true sizes.)

4. This does not necessarily mean that they were painted from impossible or inaccessible positions. *The Love Letter*, as remarked in Chapter 4, may show a different room altogether. An alternative possibility, discussed below, is that the back wall contained an opening, either a door or an internal window.

5. Wheelock, *Vermeer*, p. 148.

6. Cesare Ripa, *Iconologia*, trans. D P Pers (Amsterdam 1644).

7. The photographic reconstruction of the reflection described in Chapter 7 shows that this single patch is likely to correspond in fact to *two* fixed lights together. The image of the central post of the frame is so thin as to be invisible and the bright images of the glazed areas have spread and joined.

8. E M Gifford, 'Painting light: recent observations on Vermeer's technique', in I Gaskell and M Jonker (eds.), *Vermeer Studies* pp. 184–99.

9. Ibid., p. 187.

10. Ibid., p. 185. As Gifford says (p. 188), 'It is not clear . . . how he carried out the other function of a painted sketch: the painted lines that lay out the details of the composition.'

11. See for example *Dutch Church Painters*, catalogue of an exhibition at the National Gallery of Scotland, 6 July–9 September 1984, pp. 34–5, which show the construction lines in Pieter Saenredam's 'Great Church at Haarlem' under infrared reflectography.

12. Hyatt Mayor was alert to this problem ('The photographic eye', p. 19) and made some suggestions as to how Vermeer might have overcome it, by using a camera with a mirror, or with two lenses.

13. See C H de Jonge, *Dutch Tiles*, p. 12 and illustrations 3a and 3b.

14. See for example Pieter de Hooch, *Two Women Beside a Linen Chest*, Rijksmuseum, Amsterdam.

15. P Thornton *Seventeenth-Century Interior Decoration in England, France and Holland*, Yale University Press, New Haven, Conn. (1978), p. 141.

16. Naturally, if the 'projection screen' had been fixed in a doorway, Vermeer would have needed some separate means of getting from one room to the other, so as to move from inspecting the camera image to viewing the scene directly.

17. Zahn, *Oculus Artificialis Teledioptricus,* Vol III, unpaginated, plate following table of contents.

18. Ibid., Vol II, figure xvi, p. 231.

19. The tile pattern in *Woman with a Lute* meets the far wall in a row of three-quarter and quarter tiles. This might perhaps be a result of inaccurate repainting. The picture is in poor condition generally.

7 MORE EVIDENCE, FROM REBUILDING VERMEER'S STUDIO

1. Another possibility is that this window was shuttered or curtained.

2. One other shadow which proved impossible to reproduce satisfactorily was that cast

on the reveal of the pier by the horizontal member of the window frame, between the casements and the fixed lights. There is also a vertical strip of shadow above this in the reconstruction, cast by the frame of the fixed light, which does not occur in the painting, but this is due to a small error in the design of the model.

3. Compare the magnified figure of the soldier relative to the woman in *Officer and Laughing Girl*.

4. That is, for *The Music Lesson*, *The Concert*, *Lady Standing at the Virginals*, and *Lady Writing a Letter, with her Maid*; but not for *The Glass of Wine* or *The Girl with a Wineglass*, for the reasons explained in Chapter 6, n. 1.

5. It might have provided a clue if the sizes of the 'projected images' of these two paintings on the back wall had turned out to be in some simple ratio to the actual sizes of the canvases—say one half. That is to say, Vermeer might simply have doubled the dimensions. However, this is not the case. In fact the viewpoint of *Allegory of the Faith* is very close to the back wall, and so its projected image on the wall would have been inconveniently small to work from.

6. We can work out the depth of the doorway in the scene depicted in *The Love Letter*, albeit rather approximately, since we know it is behind—possibly just behind—the chair in the right foreground of the picture. If we assume that the seat of this chair is say 48 cm above the floor, this fixes the chair's position in plan. The consequence is that the room is at most 5.05 metres deep (16′ 8″, or eleven and a half tiles on the diagonal) as against 6.63 metres (21′ 10″, or sixteen tiles) for *The Music Lesson*. With a room 6.63 metres deep, the chair position in *The Love Letter* would be such that the seat would have to be quite improbably high to be compatible with what Vermeer shows.

In fact the more we study this doorway, the more problematic it becomes. The wall in the left foreground, on which the map hangs, is set at right angles to the opening itself. It is possible to know this from the way in which the image of the map recedes to the vanishing point. We the viewers are in the *corner* of an adjacent room, as it seems. Consider what we see of the right-hand edge of the doorway itself, above the knob on the back of the chair. The light strikes the plasterwork of the reveal from the left, to form a bright vertical stripe. But if the wall in question is parallel to the picture plane, parallel to the far wall of the room beyond, as it seems to be, and is thus viewed frontally, then this door reveal should not be visible. The vanishing point of the picture is to the *right* of the doorway. We see the opening therefore slightly from the right—so we should see the reveal on the left-hand side of the opening—but *not* the reveal on the right.

The width of the opening as calculated from the reconstruction is only 62 cm, rather narrow for a door. (Most doors are closer to 75 cm wide.) Finally the broom seen just inside the further room, which seems at first sight to lean against the far side of the wall in the foreground, cannot in fact be doing so. It is too far away from the opposite side of the wall and must be propped on some other support. There is the further matter of the fireplace, in a position where no fireplace is found in Vermeer's other comparable paintings of interiors.

7. Perhaps the matter could be decided by radiography.

8. It is conceivable that this might be tested experimentally (up to a point) despite the lack of visible tiles, given that we now have very precise dimensions for many items depicted in these paintings, including the chairs, maps, and window casements. The

viewpoints might be located through a process of establishing sightlines relative to the window embrasures, combined with calculations of distances required to achieve the observed scale reductions in maps or other objects close to the far wall. I have not, however, pursued this line of investigation.

8 ARGUMENTS AGAINST VERMEER'S USE OF THE CAMERA

1. J-L Delsaute, 'The camera obscura and painting'.

2. Ibid., p. 120.

3. Zahn, *Oculus Artificialis Teledioptricus.*

4. W J 's Gravesande, *Essai de Perspective.* The first edition (1711) has no figures. Figure 67 here is reproduced from the edition of 1717. 's Gravesande also describes a second, smaller type of camera, in the form of a three-sided box. The user, covered with a black cloth, looks into this box from the open side. A lens tube and 45° mirror are again mounted at the top of the box. The arrangement is similar to that of the camera shown in Figure 7.

5. Barbaro, *La Pratica della Perspettiva*, pp. 192–3.

6. Niceron, *La Perspective Curieuse,* pp. 21–3. Another French work, published somewhat earlier, in which the camera obscura is also described is J Leurechon, *Récréation Mathematicque. Composée de Plusieurs Problèmes Plaisant et Facétieux*, Pont à Mousson (1626). Constantijn Huygens possessed a copy (*Catalogus de Bibliotheck van Constantyn Huygens . . . 1688*, W P van Stokum, The Hague, 1903, no. 591).

7. Delsaute, 'The camera obscura and painting', p. 112.

8. A A Mills, 'Vermeer and the camera obscura: some practical considerations', *Leonardo* vol. 31, no. 3 (1998), pp. 213–18.

9. See Hammond, *The Camera Obscura, passim*. Occasional optical improvements *were* introduced, as for example in William Storer's 'Royal Accurate Delineator', patented in 1778, which had two large rectangular biconvex lenses to produce the image, plus a third lens beneath the ground glass screen (Hammond, p. 78). In the early 19th century Wollaston proposed using a meniscus lens instead of a biconvex lens, and the French optician Chevalier developed a meniscus prism which increased the area in sharp focus (Hammond, p. 122). But cameras with single biconvex lenses continued to be manufactured into the 1850s. The earliest true photographic cameras, built by Daguerre, Niépce, and Fox Talbot, were all improvised from or modelled on instruments of this type.

10. H C King, *The History of the Telescope*, Charles Griffin, High Wycombe (1955), p. 29, citing J O Halliwell, *Rara Mathematica* (1839) and D Baxandall, *Transactions of the Optical Society*, 24 (1923), p. 306.

11. Monconys, *Journal des Voyages,* part II, p. 145. Monconys gives dimensions in 'pouces'. I have assumed that 1 pouce [thumb] = 27 mm.

12. A A Mills and M L Jones, 'Three lenses by Constantine Huygens in the possession of the Royal Society of London', *Annals of Science* 46 (1989), pp. 173–82. There is perhaps

some irony in Allan Mills's claim that Vermeer could not have had access to anything larger than spectacle lenses, given the information in this paper. Mills has suggested ('Vermeer and the camera obscura', p. 216) that the 'impecunious' Vermeer would have been too poor to afford expensive telescope lenses. He has perhaps got this impression from the fact that Vermeer died in debt, but this was a consequence of Holland's war with France and the resulting difficulties of the Dutch economy in general and the art market in particular. Montias (an economic historian) shows that, on the contrary, Vermeer enjoyed for most of his life a certain financial independence, albeit a sometimes precarious one, not least because of the regular help he received from Maria Thins, a wealthy woman (*Vermeer and his Milieu*, pp. 183–6).

13. J M Montias, 'Vermeer and his milieu: conclusion of an archival study', *Oud Holland* vol. XCIV (1980), pp. 44–62. See p. 60, n. 97.

14. Christiaan Huygens, *Oeuvres Complètes*, Société Hollandaise des Sciences, The Hague (1888), vol. 1, letters 202, 233, and 234.

15. Wheelock, *Perspective, Optics and Delft Artists Around 1650*, p. 277.

16. See, for example, Wheelock, *Vermeer*, pp. 39 and 112.

17. Swillens, *Johannes Vermeer*, p. 68.

18. Ibid., p. 77.

19. Fink, 'Vermeer's use of the camera obscura', p. 504.

20. Ibid., p. 505.

21. See, for example, E Rothstein, 'Of pinpoints and cameras: the science of Vermeer's art', *New York Times*, 24 January 1996.

22. Wheelock, *Vermeer and the Art of Painting*, p. 126. Compare also D Arasse, *Vermeer: Faith in Painting*, Princeton University Press, Princeton, NJ (1994), p. 26.

23. Daniel Arasse by contrast argues that 'the orb does not reflect any of what is represented in the room. . . . Indeed, if we look closely, we cannot distinguish any figure or recognize any object visible elsewhere in the painting—except, precisely, for the window of the studio which is not visible, hidden as it is by the tapestry.' ('Vermeer's private allegories', in I Gaskell and M Jonker (eds.), *Vermeer Studies*, pp. 341–9. See pp. 344–5.) Had Arasse had the benefit of my Figure 53, perhaps he might have come to a different view.

24. Arasse, *Vermeer: Faith in Painting*, pp. 24–5. (Arasse also comments on the transformation of *The Procuress* in *Lady Seated at the Virginals*.) Again, what appears to be the same (unidentified) painting of the Finding of Moses appears both in *The Astronomer* and in *Lady Writing a Letter, with her Maid*—but at very different scales and cropped differently in the two cases.

25. Wheelock, *Vermeer*, p. 100. There is, however, more justice in another comparable observation of Wheelock's (ibid., p. 108), that the bottom member of the picture frame in *Woman Holding a Balance* has been adjusted in its alignment, to make room for the balance.

26. Wheelock, *Vermeer and the Art of Painting*, p. 94.

27. Kenneth Clark, *Landscape into Art*, Harper and Row, New York (1976), p. 65; Gowing, *Vermeer*, p. 128.

28. Wheelock, *Vermeer and the Art of Painting*, p. 77.

29. Ibid., pp. 78–9. Wheelock's position on this question has evolved. In 1977 he was inclined to reject comparisons with 18th- and 19th-century views as inconclusive, since these might themselves involve perspective distortions. 'Although the site is known, the area today is so completely transformed that an analysis cannot be made of the veracity of Vermeer's image.' (*Perspective, Optics and Delft Artists Around 1650*, p. 297).

30. For more detailed descriptions of Alberti's veil and Dürer's devices, see, for example, Kemp, *The Science of Art*, pp. 169–72. According to the 18th-century writer Jean-Baptiste Descamps, the Leiden painter Gerrit Dou (Vermeer's contemporary) invented a perspective device which combined a frame carrying a grid of threads, somewhat like Dürer's, with a concave lens to reduce the image to the precise scale of the painting (*La Vie des Peintres Flamands, Allemands et Hollandais*, 1751).

31. Wilenski, *An Introduction to Dutch Art*, pp. 280–90; M Mocquot, 'Vermeer et le portrait en double miroir', *Le Club Française de la Medaille* 18 (1968), pp. 58–69; N Konstam, 'Vermeer's method of observation', *The Artist*, vol. 95, issue 587, no. 1 (January 1980), pp. 22–5.

32. Wilenski, *An Introduction to Dutch Art*, p. 281.

33. I am inclined to think, against the conventional wisdom, that this was how Filippo Brunelleschi painted his famous (but lost) panel of the Baptistery in Florence, reputedly the first work ever made in correct linear perspective. Such a technique would account for the silvering of the sky, if the support was actually a metal mirror, and this part had been left unpainted. It would also provide an additional reason for viewing the picture—as Brunelleschi intended—in a (second) mirror, since a painting that is traced over a mirror is itself mirrored relative to the real scene.

34. To show that the image of one's head in a mirror has only half the dimensions of the real head, try this experiment. Stand in front of a mirror and close one eye. Put a ruler on the image of your head. The image will measure about 6 cm across, where in reality the head is around 12 cm wide.

Konstam, 'Vermeer's method of observation', confines his attention to *Allegory of Painting*, where he imagines that the figure of the artist is a self-portrait (if so, one of the most modest in the history of art). According to Konstam, Vermeer would thus have needed two mirrors to see himself from the back—a tortuous argument indeed.

35. Swillens, *Johannes Vermeer*, p. 143.

36. Ibid., p. 77.

37. H Vredeman de Vries, *Perspective*, The Hague and Leiden (1604–5); H Hondius, *Institutio Artis Perspectivae*, The Hague (1622); S Marolois, *Opera Mathematica, ou Oeuvres Mathematiques Traictons de Geometrie, Perspective, Architecture et Fortification*, The Hague (1614). Other editions and works—and other authors—could be cited.

38. J Wadum, 'Johannes Vermeer's use of perspective', in A Wallert, E Hermens, and M Peek (eds.), *Historical Painting Techniques, Materials and Studio Practice*, preprints of a

symposium held at the University of Leiden, Santa Monica (1995), pp. 148–54; and 'Vermeer in perspective', in A K Wheelock (ed.), *Johannes Vermeer*, pp. 67–79.

39. J Wadum, 'Vermeer and spatial illusion' in *The Scholarly World of Vermeer*, Museum van het Boek/Museum Meermanno-Westreenianum, The Hague; Waanders Publishers, Zwolle (1996), pp. 30–49. See p. 48.

40. K G Hultén, 'Zu Vermeer's atelierbild', *Konsthistorisk Tidskrift* 18 (1949), pp. 90–8.

41. See Wadum, 'Vermeer in perspective' p. 68.

42. Ibid., p. 68. As Wadum notes, the distance points, on the other hand, lie always outside the area of the canvas. He speculates that Vermeer placed his paintings or drawings against a large board, or a wall, in which he could put pins to mark the distance points.

43. Wadum in 'Vermeer in perspective' says, for example, that there is no remaining trace on the sides of Vermeer's canvases of the so-called 'height wall' sometimes used to calculate the heights of points and lines in the picture (p. 70).

44. Wadum, 'Vermeer and spatial illusion'. See p. 47.

45. Compare, for example, the diagrams reproduced by Kemp, *The Science of Art*, pp. 110–12.

46. Swillens agrees (*Johannes Vermeer*, p. 75). This has not stopped some critics from trying to find meanings. The woman's elbow in *The Music Lesson*, for example, is the joint at which her arm must move to make the music which is the subject of the painting. Wheelock (*Vermeer*, p. 82) says that the vanishing point of *Officer and Laughing Girl* is 'midway between the eyes of the soldier and the woman'. It *is* approximately on this line, but it is not 'midway': it is close to the brim of the officer's hat. The only other case, besides *Lady Writing a Letter, with her Maid*, in which the positioning of the vanishing point is possibly significant, in my estimation, is in *The Girl with a Wineglass*, where it is found on the head of the gloomy gentleman seated in the background—but even this falls in the man's *hair*.

9 THE INFLUENCE OF THE CAMERA ON VERMEER'S PAINTING STYLE

1. Swillens, *Johannes Vermeer*, p. 120.

2. G Q Williams, 'Projected actuality', *British Journal of Aesthetics*, vol. 35 (July 1995), pp. 273–7. See p. 275.

3. G Q Williams, personal communication.

4. Gowing, *Vermeer*, p. 46.

5. André Malraux, *Les Voix du Silence* (1951), trans. Stuart Gilbert as *The Voices of Silence*, Secker and Warburg, London (1954), p. 478.

6. R Huyghe, *Vermeer et Proust* (1936), quoted in Blankert, Montias, and Aillaud, *Vermeer*, p. 217.

7. Gowing, *Vermeer*, p. 25.

8. Gowing, *Vermeer*, p. 117.

9. G Q Williams, personal communication.

10. Gowing, *Vermeer*, p. 23.

11. Ibid., p. 19.

12. Clark, *Landscape into Art*, p. 65.

13. P Claudel, *La Revue de Paris*, 1 et 15 février 1935, quoted in Schwarz, 'Vermeer and the camera obscura', p. 171 [*author's translation*].

14. Kemp, *The Science of Art*, p. 193.

15. G Q Williams, 'Projected actuality', p. 274.

16. Malraux, *Les Voix du Silence*, p. 476.

17. Montias, *Vermeer and his Milieu*, p. 160.

18. See A Scharf, *Art and Photography*, Penguin, Harmondsworth (2nd edn 1974), p. 240, plate 175: Richard Polak, *The Painter and his Model*, 1915. Polak published his pictures in *Photographs from Life in Old Dutch Costume*, 1914.

19. Blankert, *Vermeer of Delft*, p. 33.

20. C de Tolnay, '*L'Atelier* de Vermeer', *Gazette des Beaux-Arts*, vol. XLI (1953), pp. 265–72. English translation by de Tolnay, pp. 292–4. See p. 292.

21. Gowing, *Vermeer*, p. 26.

FURTHER READING

Alpers S, *The Art of Describing: Dutch Art in the Seventeenth Century*, John Murray, London (1983)

Bell A E, *Christian Huygens and the Development of Science in the Seventeenth Century*, Edward Arnold, London (1947)

Blankert A with R Ruurs and W L van der Watering, *Vermeer of Delft*, Phaidon, Oxford (1978)

Brown C, *Carel Fabritius*, Phaidon, Oxford (1981)

Brusati C, *Artifice and Illusion: The Art and Writing of Samuel Van Hoogstraten*, Chicago University Press, Chicago (1995)

Colie R L, *'Some Thankfulnesse to Constantine': A Study of English Influence upon the Early Works of Constantijn Huygens*, Nijhoff, The Hague (1956)

Delsaute J-L, 'The camera obscura and painting in the sixteenth and seventeenth centuries', in I Gaskell and M Jonker (eds.), *Vermeer Studies*, National Gallery of Art, Washington and Yale University Press, New Haven, Conn., pp. 111–23 (1998)

Dobell C, *Antony van Leeuwenhoek and his 'Little Animals'*, John Bale, Sons and Danielsson, London (1932)

Fink D A, 'Vermeer's use of the camera obscura—a comparative study', *Art Bulletin* 53, pp. 493–505 (1971)

Gaskell I and M Jonker (eds.), *Vermeer Studies*, National Gallery of Art, Washington and Yale University Press, New Haven, Conn. (1998)

Gernsheim H with A Gernsheim, *The History of Photography: From the Camera Obscura to the Beginning of the Modern Era*, Thames and Hudson, London (1955, rev. 1969)

Gowing L, *Vermeer*, Faber, London. (1952; 2nd edn 1970)

Hammond J H, *The Camera Obscura: A Chronicle*, Adam Hilger, Bristol (1981)

Hyatt Mayor A, 'The photographic eye', *Bulletin of the Metropolitan Museum of Art*, new series vol. V, no. 1, pp. 15–26 (1946)

Kemp M J, *The Science of Art: Optical Themes in Western Art from Brunelleschi to Seurat*, Yale University Press, New Haven, Conn. (1990)

Lindberg D C, 'The theory of pinhole images from antiquity to the thirteenth century', *Archive for History of the Exact Sciences* 5, pp. 154–76 (1968–9)

Maarseveen M P van, *Vermeer of Delft: His Life and Times*, Stedelijk Museum het Prinsenhof, Delft and Bekking Publishers, Amersfoort (1996)

Mills A A, 'Vermeer and the camera obscura: some practical considerations', *Leonardo*, vol. 31, no. 3 (1998), pp. 213–18

Montias J M, *Vermeer and his Milieu: A Web of Social History*, Princeton University Press, Princeton, NJ (1989)

Pennell J, 'Photography as a hindrance and a help to art', *British Journal of Photography*, no. 1618, vol. XXXVIII (1891), pp. 294–6

Potonniée G, *Histoire de la Découverte de la Photographie*, Paris (1925), trans. E Epstean as *The History of the Discovery of Photography*, Tennant and Ward, New York (1936)

Ruestow E G, *The Microscope in the Dutch Republic: The Shaping of Discovery*, Cambridge University Press, Cambridge (1996)

Schierbeek A, *Measuring the Invisible World: The Life and Works of Antoni van Leeuwenhoek FRS*, Abelard-Schumann, London (1959)

Schwarz H, 'Vermeer and the camera obscura', *Pantheon* XXIV (May–June 1966), pp. 170–82

Seymour Jr C, 'Dark chamber and light-filled room: Vermeer and the camera obscura', *Art Bulletin* 46 (1964), pp. 323–31

Swillens P T A, *Johannes Vermeer: Painter of Delft 1632–1675*, Spectrum, Utrecht and Brussels (1950)

Wadum J, 'Vermeer in perspective', in A K Wheelock (ed.) *Johannes Vermeer*, catalogue of an exhibition held at the National Gallery of Art, Washington and the Mauritshuis, The Hague; Yale University Press, New Haven, Conn. (1995–6), pp. 67–79

Waterhouse J, 'Notes on the early history of the camera obscura', *The Photographic Journal*, vol. XXV, 9 (31 May 1901), pp. 270–90

Welu J A, 'Vermeer: his cartographic sources', *Art Bulletin* 57 (1975), pp. 529–47

Wheelock A K, *Perspective, Optics and Delft Artists Around 1650* (reprint of dissertation submitted to Harvard University 1973), Garland, New York (1977)

Wheelock A K, *Vermeer and the Art of Painting*, Yale University Press, New Haven, Conn. (1995)

Wheelock A K (ed.), *Johannes Vermeer*, catalogue of an exhibition held at the National Gallery of Art, Washington and the Mauritshuis, The Hague; Yale University Press, New Haven, Conn. (1995–6)

PICTURE CREDITS

The publishers would like to thank the following individuals and institutions who have kindly given permission to reproduce the illustrations listed below.

Figures

1. Cambridge University Library.
2. © Philip Steadman.
3. © Philip Steadman.
4. NMPFT/Science & Society Picture Library.
5. NMPFT/Science & Society Picture Library.
6. © Philip Steadman.
7. The British Museum.
8. NMPFT/Science & Society Picture Library.
9. The Royal Collection © 2000, Her Majesty Queen Elizabeth II
10. Rijksmuseum, Amsterdam.
11. Private Collection.
12. Photograph, through a camera obscura, simulating detail from *Girl with a Red Hat*. Photograph by Henry Beville, published in Seymour, *Dark Chamber and Light-Filled Room*, 1964.
13. Detail from Willem Blaeu's pictorial plan of Delft, 1648, reproduced in P T A Swillens, *Johannes Vermeer, Painter of Delft 1632–1675*, Spectrum, Utrecht/Brussels.
14. From a Private Collection.
15. Westfries Museum, Holland.
16. The Bridgeman Art Library/Mauritshuis, The Hague.
17. © Photo RMN/C Jean.
18. Rijksmuseum, Amsterdam.
19. Rijksmuseum, Amsterdam.
20. Detail of 1675–78 Plan of Delft, from H L Houtzager et al, *De Kaart Figuratief van Delft*, Elmar B V, Rijswijck (1997).

21. Delft Municipal Archive.
22. The Royal Collection © 2000, Her Majesty Queen Elizabeth II.
23. © Philip Steadman.
24. © Philip Steadman.
25. Rijksmuseum, Amsterdam.
26. © Philip Steadman.
27. © Philip Steadman.
28. © Philip Steadman.
29. © Philip Steadman.
30. © Philip Steadman.
31. © Philip Steadman.
32. © Philip Steadman.
33. © Philip Steadman.
34. Reproduced in P T A Swillens, *Johannes Vermeer, Painter of Delft 1632–1675*, Spectrum, Utrecht/Brussels.
35. © Philip Steadman.
36. © National Gallery, London (left); © Philip Steadman (right).
37. Philip Steadman.
38. The Royal Collection © 2000, Her Majesty Queen Elizabeth II (left); © Philip Steadman (right).
39. © Philip Steadman.
40. © Philip Steadman.
41. NMPFT/Science & Society Picture Library.
42. Rijksmuseum, Amsterdam.
43. British Library, London.
44. Bibliothèque Nationale, Paris.
45. Courtesy, Museum of Fine Arts, Boston. Reproduced with permission © 2000 Museum of Fine Arts, Boston. All rights reserved.
46. © Philip Steadman.
47. © Philip Steadman.
48. Cartography Department, Technical University, Delft.
49. © Philip Steadman.
50. © Philip Steadman.
51. © Philip Steadman.
52. © Philip Steadman.
53. The Metropolitan Museum of Art, Washington (top left); © Philip Steadman (right and below).
54. NMPFT/Science & Society Picture Library.
55. © Philip Steadman.

56. © National Gallery, London.
57. © Philip Steadman.
58. © Philip Steadman.
59. © Philip Steadman.
60. © Philip Steadman.
61. © Philip Steadman.
62. ©Philip Steadman (left); bpk, Berlin/Gemäldegalerie/ Jorg P Anders (right).
63. © Philip Steadman (left); Herzog Anton Ulrich-Museum/ Bernd-Peter Kelser (right).
64. © Philip Steadman (left); National Gallery of Ireland, Dublin (right).
65. © Philip Steadman.
66. © Philip Steadman.
67. Science & Society Picture Library.
68. Photo © Allan Mills.
69. bpk, Berlin/Gemäldegalerie/Jorg P Anders.
70. © Philip Steadman.
71. © Philip Steadman.
72. (a) Herzog Anton Ulrich-Museum/Bernd-Peter Kelser; (b) Andrew Mellon Collection, © 2000 Board of Trustees, National Gallery of Art, Washington; (c) Kunsthistorisches Museum, Vienna.

Plates

1. Copyright The Frick Collection, New York.
2. Andrew Mellon Collection, © 2000 Board of Trustees, National Gallery of Art, Washington DC.
3. The Bridgeman Art Library/Mauritshuis, The Hague.
4. Städelsches Kunstinstitut, Frankfurt/Jochen Beyer.
5. © Philip Steadman.
6. The Royal Collection © 2000, Her Majesty Queen Elizabeth II.
7. © Philip Steadman.
8. The Bridgeman Art Library/Isabella Stewart Gardner Museum, Boston.
9. © Philip Steadman.
10. © National Gallery, London.

The publisher and the author apologize for any errors or omissions in the above list. If contacted they will be pleased to rectify these at the earliest opportunity.

INDEX

Numbers in italic indicate illustrations.